A NOTE ON THE AUTHOR

Liam Drew is a writer, former neurobiologist and mammal. He has a PhD in sensory biology from University College London and spent 12 years researching schizophrenia, pain and the birth of new neurons in the adult mammalian brain at Columbia University, New York and at UCL. His writing has appeared in *Nature, New Scientist, Slate* and the *Guardian*. He lives in Kent with his wife and two daughters.

Also available in the Bloomsbury Sigma series:

I, MAMMAL

THE STORY OF WHAT
MAKES US MAMMALS

Liam Drew

BLOOMSBURY SIGMA
LONDON · OXFORD · NEW YORK · NEW DELHI · SYDNEY

BLOOMSBURY SIGMA
Bloomsbury Publishing Plc
50 Bedford Square, London, WC1B 3DP, UK

BLOOMSBURY, BLOOMSBURY SIGMA and the Bloomsbury Sigma logo
are trademarks of Bloomsbury Publishing Plc

First published in the United Kingdom in 2017
This edition published 2019

A catalogue record for this book is available from the British Library.

Library of Congress Cataloguing-in-Publication data has been applied for.

ISBN: PB: 978-1-4729-2291-5; ebook: 978-1-4729-2292-2

2 4 6 8 10 9 7 5 3 1

Illustrations by Marc Dando (unless credited otherwise)

Typeset by Deanta Global Publishing Services, Chennai, India
Printed and bound in Great Britain by CPI Group (UK) Ltd, Croydon CR0 4YY

To find out more about our authors and books visit www.bloomsbury.com.
and sign up for our newsletters.

For Mariana, Isabella and Cristina.

And for Cliff

Contents

Of animals, some resemble one another in all their parts, while others have parts wherein they differ.

Aristotle

My Family and Other Mammals

In the standard modern fashion, Cristina went to the bathroom, and I paced the living room. That month had been our fifth attempt to fall pregnant but the first that Cristina's clockwork body had suggested it was worthwhile taking a plastic strip from the box; a strip that would tell us if a nascent placenta was already releasing a stream of hormones into its mother's blood.

Seven months later, eight weeks early, Isabella was born. The morning after, I sat at Cristina's bedside exhausted and shocked. Isabella was too immature to suckle, and we were uncertain if her mother's body was ready to lactate. I watched for 20 minutes as Cristina compressed and released a handheld plastic pump, until, finally, with a tiny plastic syringe, she collected a few droplets of colostrum from around her nipples. In the ward where Isabella slept, our nurse, whose scrubs we came to suspect concealed a pair of angel wings, held aloft that syringe and reacted as if Cristina were the finest dairy cow at the county fair. He added those drops of new milk to the formula he'd prepared and gently pushed the mixture through a plastic tube that passed from Isabella's mouth to her stomach.

Soon, that handheld pump was replaced by a heavy electrical contraption, and as Isabella stayed in hospital, our nights at home those first weeks were broken not by newborn wailing but by the mechanical whirring of Cristina's determination. Only when Isabella was a month old did a nurse called Joy suddenly say that it was time she tried a bottle. Tense with anticipation, we watched Joy move the rubber teat toward Isabella's tube-free mouth – then we gasped, beamed, and felt a knot of anxiety unravel as Isabella suckled for the first time.

Cristina now took to circling a nipple before Isabella, and, after another week, Isabella latched on. It was a moment of pure jubilation. Flushed with awe, happiness and relief, I sat entranced by the rhythmic formation and relaxation of a dimple in my daughter's cheek.

I did not, at that point, think, 'Look at my partner and daughter engaging in a uniquely mammalian activity.' Certainly, I knew this was the case, but I'm not sure that I did a whole lot of thinking in those weeks, not in any protracted or intellectual sense. I merely reacted and responded. And felt. I felt things very acutely. I was using unfamiliar parts of my brain. I was powered by something new.

Isabella was in hospital for two months, the first of which was the most gruelling of both Cristina's life and mine. But it was also joyous. The neonatal intensive care unit continually swung us between moments of blissful parental contentment and black wells of helpless what-if …? These were fears of a new magnitude and fears that seemed to feed on the energy of the happiness they trampled.

As often as possible, we'd remove Isabella from her temperature-controlled crib and rest her on our chests, Cristina's usually. 'Kangaroo Care' they called it. But most of my time at the NICU, I sat quietly beside the crib as my daughter slept. I'd talk a little, insert a sterilised hand through the plastic portal to ensure she knew I was there, and – with all my strength – will her well.

I pleaded for Isabella's body to do – where it was – what it should have done inside a womb. My focus was whatever the doctors had been most concerned about that day. Sometimes it was her GI tract, other times it was her breathing or feeding … I would cast in my mind a picture of her lungs, say, and the nerves that should have been controlling them, and implore those structures to develop as they should, for the nerves to reach out and grip their intended targets.

The day before what had once been Isabella's due date we departed the hospital. The three of us descended a lift shaft as if it were a second birth canal expelling us together into the world. Isabella was fragile, but she was healthy. We were lucky.

I am now a more grateful person, but the fall-out was greater than that. I witnessed too the physical toll taken by pregnancy, birth and breastfeeding. The two of us, Cristina especially, waded through previously unknown sleeplessness and exhaustion. I watched her become a mother mentally as well as physically, as I too experienced fatherhood hijack my psyche. I changed. Before, I'd seen myself primarily as a free-floating cerebrum, a mind, a stream of cognition: *I had thought, therefore I had been*. And now it was different. For 20 years, I'd studied biology; finally, I understood that I *was* biology.

The first chapter of this book is an investigation into why the vast majority of male mammals carry their testicles in wrinkled carry cases beyond the safety of their abdomens, and it was written as a standalone article before Cristina and I embarked upon becoming parents. (Indeed, after each failure to become pregnant, we wondered about the errant football that's painfully felt impact had inspired that story.) When *Slate* published the article after Isabella's birth, I thought that was the end of it. But then – after months of obsessing over feeding Isabella – I found myself plotting a similar survey of what evolutionary biology had to say about the origins of lactation.

Milk, like the scrotum, was a biological peculiarity seen only in mammals. No other type of animal feeds its young as mammals do; indeed, mammary glands inspired the very name mammal. The terrain felt familiar, a theme was emerging – I was back inside ancient mammalian history considering how, aeons ago, a very particular type of animal had evolved a very particular new trait, a trait that shaped the way I now lived. And thinking more widely about the things that had preoccupied me since becoming a father, many of them, I saw, had been quintessentially mammalian. Our daughter had developed inside a womb, nourished by a placenta. In the hospital, her temperature had been carefully monitored to ensure that she remained warmer than her surroundings. And weren't the emotional upheavals of becoming parents also pretty mammalian things to have happened? Certainly, the brain that had driven this transition – that had wrestled with attachment and anxiety, and that had rendered everything as an experiential reality – was covered in a folded sheet of grey matter that existed only in mammals.

Before, when I'd considered what I owed evolution, my interest had always lain with this cerebrum of mine, and how apes might have crossed a mostly brain-defined threshold to become human. But now, pushed by the animalistic urgency of parenthood, I wanted to go further back. I wanted to put all these pieces together and attempt to understand what it was that made me a mammal.

Mammal species of the world

Aardvarks, aardwolves and alpacas. Beavers, coypus and dik-diks. Elephants, foxes, giraffes, hyenas, impalas, jackals, kangaroos, leopards, manatees and narwhals. Opossums and possums. Quokkas and rhinoceroses. Squirrels and tapirs. The Uganda kob, voles and wildebeests. Xenarthra (it's a group of mammals – the most valid X I have – we'll get to it). Yaks and zebras.[*]

Mammal is a word that binds 150 tonnes of blue whale to two grams of Etruscan shrew, and an inch-long bumblebee bat to six tonnes of African elephant; it connects a tiger's poise to a mole's tunnelling and a kangaroo's bounce, and it bonds the peculiarity of an armadillo to the familiarity of a cat.

Mammals live in all the habitats of this Earth. They gallop, bound, amble, burrow, glide, swim and fly. So widely have

---

[*] An aardwolf is a hyena-like creature, and a dik-dik is a small antelope. A narwhal is a whale with a unicorn's horn. A quokka is a marsupial about the size of a cat that resembles a cross between a kangaroo, a mouse and a rabbit. The Uganda kob is another type of antelope, the males of which mark their territories by whistling.

they proliferated, biologists frequently call the period since the demise of the dinosaurs 'The Age of Mammals'.

According to the third edition of *Mammal Species of the World: A Taxonomic and Geographic Reference*, published in 2005 – a two-volume work edited by Don Wilson and DeeAnn Reeder – there are currently 5,416 species of mammals.

And the upcoming fourth edition of *MSW* will include more. Since the third edition, zoologists have discovered a new monkey species in the Democratic Republic of the Congo, new blossom bats in Papua New Guinea, a new dolphin in Australia, novel mice in Cyprus, Indonesian toothless rats, the Star Wars gibbon in China, and a previously unidentified species of leopard in Malaysia.[*]

The twinned volumes of *Mammal Species of the World* that I consulted stood squat on the shelf. Labels stuck to their front covers read 'NOT TO BE REMOVED FROM THE LIBRARY'. They were heavy, authoritative objects that I was glad to physically interact with, rather than clicking through a website. When I put down the 1,400-page second volume, it made the table shudder.

The tome is essentially just a long and magnificent list. Each standard format entry gives the species' scientific and common names, followed by the person who first described and/or named that mammal, and the year in which this was done. Then, in somewhere between 5 and 20 lines, it gives the hard, taxonomic facts of the species' existence. There are no pictures and no actual descriptions of the animals. The book's vastness is testament to both biology's variety and the human endeavour that goes into describing it, yet neither of these things is flaunted; the reader is trusted to have the imagination to do these facts justice.

At the highest level, mammals are divided into three unevenly sized groups: the monotremes, the marsupials and the placental mammals. *MSW* begins with the monotremes

---

[*] Sometimes species are entirely new, and sometimes it is decided that what were previously considered two subspecies are, in fact, distinct enough to be called two species.

– Australasia's duck-billed platypus and four species of closely related spiny anteaters, or echidnas (pronounced e-KID-nas). Next, there are the marsupials, of which there are 331 species listed in *MSW*. Most marsupials live alongside the monotremes in Australasia, but a good number inhabit South America and one species, the Virginia opossum, has made a home of North America.

This leaves 5,080 species of globe-spanning placental mammals. Of these, 1,116 are mammaldom's aviators, the bats, and 2,277 are rodents. In fact, rodents take up the entire second volume of *MSW*. Rats, mice, voles, squirrels, chipmunks, gerbils, guinea pigs and their kin account for more than 40 per cent of all mammals.

Non-rodent, non-bat placental mammals number, therefore, just 1,687. Among these are those species that leave a visitor to the Serengeti breathless: the swarming, perpetually migrating herds of wildebeest, and the gazelles and zebras that travel with them in search of rain; giraffes that eat tree-tops, as the grazers pluck at the grass; the cheetah that surveys who she might chase down; the roving hyenas scouting for carrion; and the pride of lions slumbering, digesting yesterday's kill. Elsewhere are the furless grey giants: herds of African elephants, bloats of hippos and crashes of rhinoceroses.

Rhinos and elephants live in Asia too. This is where rhinos' deep ancestry lies, while the elephant originated in Africa.

Mammals evolved on land – this is important – but twice, different lineages resumed fully aquatic lives. The separate ancestors of cetaceans – the dolphins and whales – and of manatees and dugongs have yielded animals who can live only in water. Seals, sea lions and walruses have also headed in this direction: they are functional on land but infinitely more graceful when swimming. The blue whale is the largest animal ever to have existed – its tongue weighs as much as an elephant – and it subsists on a diet of tiny shrimp-like krill. I've never known which species of whale swam alongside my boat off the Pacific coast of Mexico when I visited there in my early twenties, but I've never forgotten how humbled I felt seeing them. The manatee, or sea cow, appears around Floridian and Caribbean coasts and is as lovely as it is serene. This animal inspired the myths of mermaids and controls its buoyancy by adjusting its flatulence. The polar bear is another aquatically honed mammal, preying mainly on the seals that have also made the Arctic Circle their home.

Black bears will now rummage through North American rubbish bins, one of the many mammals affected directly or indirectly by humans. Descendants of wolves and feline predators live in human abodes. Sheep, cattle, pigs and goats are bred and fed until they're good to eat. The milk of cows – and of goats and sheep – is big business.

Each continent and country has its unique complement of mammalian residents. Our daughter Isabella was born in the US, a country that hosts around 500 mammal species, but the fauna of my home island of Britain – to which we've now returned – is relatively restricted. Around 100 mammal species live in Great Britain. Foxes trot. Badgers and hedgehogs amble. We have our deer, our fecund rabbits and the occasional otter. Red squirrels are an increasingly rare sight, as grey ones – Victorian imports from North America – abound. Seals and dolphins swim off our coasts, and certain whales will visit. There's little to do you harm in the UK and much to gently charm you.

Among the more eccentric of mammals worldwide is the hero shrew of Africa's Congo Basin, which has a fused backbone. One such shrew once bore the weight of a grown man balancing on one foot on its back, before scuttling off, in the words of a witness, 'none the worse for this mad experience'. Across central and southern Africa, aardvarks hoover up as many as 50,000 insects in a night through their long snouts, but annually make an exception to feast on 'aardvark cucumbers', a fruit that grows underground and that depends entirely on these animals for its propagation. A squirrel's ankle can turn through 180 degrees. Naked mole rats live for decades, barely ageing and never getting cancer. They reside in subterranean communes that are reminiscent of social insects, organised in service of a queen that can bear more than 30 pups in one litter.

On page 182 of *MSW* lies *Homo sapiens*. Our distribution is given as 'cosmopolitan', and our conservation status as 'absolutely not endangered'. Because every species has a 'type' – an original specimen to which other animals can be compared – and because *Homo sapiens* was a name the taxonomist Carl Linnaeus gave us in 1758, the type specimen for humans is stated as being in his home town: Uppsala, Sweden. Humans are summarised in 19 lines. We are given no special treatment. *Homo sapiens* is just an entry towards the end of the primate section.

Mammals have spilled out in thousands of directions, and we humans constitute just one twig of this branch of life. It might only be from this twig that a hand reaches out to draw the rest of the tree, and only here that minds formulate ideas as to how this tree grew, but still, we are very much a twig of this particular branch. We often celebrate nature for its creativity and diversity – for having produced giraffes, hero shrews and blue whales – but, ultimately, nature is conservative. If it creates a good thing, it holds on to it. When our ancestors acquired the attributes that make humans unique, they surrendered none of the traits that make us mammals.

This book is about those traits, the various characteristics that define a mammalian life, such as a human one; the glue that holds together the pages of *Mammal Species of the World*.

It is not, though, intended as an exercise in chauvinism. This is not a quasi-Victorian attempt to show how mammals, then humans, are the crowning glory of evolution. If 5,416 species seem a lot, know that there are over 10,000 species of birds and a similar number of non-avian reptiles. Add in amphibians and fishes, and there are about 66,000 species of vertebrates. Then there are about 1.3 million known invertebrates – mostly insects. Of approximately 33 basic types of animals, vertebrates are but one. So, we mammals represent just a single mode of living on this planet, and it's the characteristics that define this particular way of being that I'd like to understand a little better.

Mammalia

Mammals have existed as a recognised group since 1758. (That they've actually existed for about 210 million years, we'll get back to.) This is when Carl Linnaeus produced the 10th edition of his *Systema Naturae*, in which he christened people *Homo sapiens*.

Systema Naturae began life in 1735 as 11 large pages on which an ambitious young Swedish botanist and taxonomist attempted to list every earthly plant, animal and mineral: 'the whole of the productions and inhabitants of this our globe'. Linnaeus then updated *Systema* throughout his life, assembling 12 editions before his death, the last – finished in 1768 – containing 2,400 pages.

But the 10th edition is the most notable. Linnaeus used this version to start naming animals the way he'd named plants, using the two-name – or *binomial* – system we still use today. The first name, for example *Homo*, is the genus, or group, to which the species belongs; the second, for example *sapiens*, describes the species. *Homo sapiens* is Latin for 'wise man'. The new name was necessary because Linnaeus boldly classified people as merely another species of animal.*

Also in this edition, Linnaeus divided the animal kingdom differently from the way he had before. Regarding mammals, his decisive move was to transform whales and dolphins from fish – as they had been for nine editions – into companions of

---

* In private, Linnaeus was even more forthright in seeing humans as animals, writing to a colleague in 1747, 'I ask you and the whole world for a generic differentia between man and ape which conforms to the principles of natural history. I certainly know of none.' He later complained that to call humans apes would, however, 'bring down all the theologians on my head'.

mice, horses, and the species he was renaming *Homo sapiens*. It's hard to see why this took so long to happen. Since Aristotle, people had noted how similar whales and dolphins were to furry, warm-blooded land animals: their internal anatomies were strikingly alike, plus they breathed air, birthed live young and nursed their offspring. In fact, John Ray, the most influential taxonomist before Linnaeus, had proposed in 1692 to group whales and dolphins with other mammals, but the idea hadn't caught on. Linnaeus's reclassification, however, stuck. A collection of 184 animals clearly similar to one another and obviously distinct from all other types of animal had been successfully united and classified.[*]

This new group needed a name, however. Previously, non-aquatic mammals had been called *Quadrupedia*, or 'the four-footed ones'. This term had already been problematic given that bats, people and seals did not have four feet, and that monkeys were considered to have four hands. And now with whales and dolphins as the prominent new inclusions, an emphasis on four legs was entirely inappropriate.

The titles of Linnaeus's five other groups followed no particular rules: fish and birds were *Pisces* and *Aves*, which were old Latin names; 'amphibians' (which included what we now call reptiles) was derived from 'amphibious', describing the dual terrestrial and aquatic lifestyles of these animals; insects came from 'in sections'; and the miscellaneous category of worms, molluscs and other invertebrates was named *Vermes*, the Latin word for worm

John Ray had called his grouping, not unpleasantly, *Vivipara*, as the animals all gave birth to live young, but Linnaeus had other ideas. For his new group, the Swede derived a name from another of its members' defining characteristics. He gave no reason for this, included no

---

[*] Most of the 184 species were European residents. Amusingly, Linnaeus had believed he was close to listing all species and that the generation to follow his would probably complete the task.

justificatory note: 'Mammalia,' he simply stated, 'these and no other animals have mammae.'*

Systema Naturae prefaced each of its classes of animals, plants and minerals with a list of the features that distinguished the group. For mammals, the list read: a four-chambered heart, lungs, covered jaw, teats, five sense organs, a covering of hair ('scanty in warm climates, hardly any on aquatics'), four feet ('except in aquatics'), and 'in most a tail. Walks on the earth and speaks.'

Perhaps, then, there's something Linnaean in the approach I've taken to investigating mammalian-ness. After penning a natural history of the scrotum and contemplating a similar exploration of lactation, I started to make a quasi-Linnaean list of features unique to mammals. Some idiosyncrasies, such as being warm-blooded animals insulated by hair, are well known, while others were a surprise – we mammals, for example, have bony palates that separate our nasal cavities from our mouths, giving us the rare ability to eat and breathe at the same time.

Some items on my list became chapters in their own right, while others were grouped together to form another section. Finally, the backbone of this book became: scrotums, mammalian X and Y chromosomes, reproductive organs,

---

* The notion of a mammal has remained essentially stable since Linnaeus defined it, but it took decades for this new name to become commonplace, with references to quadrupeds remaining popular through the early 19th century. Additionally, the inconsistency of Linnaeus's group names was considered problematic, and there were attempts to rebrand mammals. In 1816, Henri de Blainville, an influential French naturalist who helped to establish that amphibians and reptiles were distinct groups, proposed that mammals be called *Pilifera* for having hair and that birds and reptiles be named *Pennifera* and *Squammifera* after their feathers and scales, respectively. John Hunter, who we'll meet in Chapter 6, liked the name *Tetracoilia* for 'four-chambered heart', but this is a feature mammals share with birds.

placentas, mammary glands, parenting, skeletons and teeth, hair-enabled warm-bloodedness, sensory capabilities, and last, large, cerebral-cortex-coated brains. Not every mammal possesses all of these, which we'll discuss as we go. But together I hope they capture the essence of mammalian biology.

I thought, initially, that presenting these traits in the order in which they had evolved would be sensible. However – as we'll also discuss – this turned out to be naive: the emergence of mammals wasn't a process of sequentially adding new traits to a pre-mammalian ancestor. Instead, given that this endeavour began with an exploration of dangling male gonads and was cemented by the birth of my first daughter, I've arranged them – from externalised sperm production through the long and nuanced reproductive biology of mammals to the nature of our adult bodies and brains – so that they roughly trace the arc of a mammalian life.

Then, with each trait my tactic has been not so different from how Isabella and now Mariana – her younger sister – approach the world: 'Why?' they ask of so many basic things, 'Why do cows make milk?', 'Why do we have legs?' For although maturity may teach us that there's not always a definitive answer to such queries – and that we might also do well to ask the more prosaic questions of 'How?', 'What?' and 'When?' – 'Why?' remains our favourite question.

'All true classification is genealogical'

Linnaeus was apparently fond of saying, 'God created, Linnaeus organised' – he did not lack self-belief. But although *Systema Naturae* was collated in a time when nature was seen as the handiwork of a higher being, across Linnaeus's lifetime, the possibility that living forms evolved over time was increasingly debated. In later life, even Linnaeus himself pondered the extent of divine creation, eventually questioning whether the genus – elephants, say – had been God's creation, while new species – the African and Asian iterations – might have arisen organically.

Then, as science developed apace, and as European explorers introduced the Old World to more and more of Earth's biological diversity, speculation about species' malleability intensified. It was a debate that would culminate 101 years after mammals were first classified, in the publication of *On the Origin of Species* – a book that struck science like a sledgehammer and that definitively split biology into pre- and post-Darwinian epochs.

The impact of Charles Darwin's great book was twofold: first, Darwin convinced the world – more than any predecessor had – that life forms do, indeed, evolve over geological time. The evidence he'd accrued to support this conjecture was overwhelming. And, second, Darwin offered a convincing explanation of why evolution happened. His theory of natural selection said that heritable traits that made organisms better able to survive and reproduce in a given environment would be passed on to more offspring than were left by organisms less well suited to their circumstances. That way useful traits proliferated at the expense of less adaptive ones, and over countless generations organisms could profoundly change form.

Additionally, Darwin later discussed sexual selection, showing that what males and females found desirable in potential breeding partners also shaped which characteristics passed down through the generations.

With that book, the world was no longer filled with timeless plants and animals doing what they'd always done. Instead, there was a story to be told. Now, if you flicked through an imaginary age-old family album, you wouldn't watch subtly different humans appearing all the way back to the start. Instead, you'd see humans morph back to having increasingly ape-like features, then monkey-like characteristics and so on, through myriad ancient mammals until you were watching amphibians, then fish and their predecessors flash by. If such an album existed, just how big it would be and how quickly you'd have to flick the pages, one can only attempt to imagine. A table wouldn't shudder beneath it; it would collapse.

For such a book to chart the whole shebang of how mammals came to exist, it would need to stretch back around 3.7 billion years to the genesis of life. And from there, there would be 2.5 billion years of biology consisting of only single-celled organisms. Somewhere mid-to-late in that single-celled history, complex cells resembling those that would later make up plant and animal bodies are born. En route to mammals, some of these complex cells started to club together, probably some 800 to 600 million years ago, to form the first multicellular animals, and by about 525 million years ago the first vertebrates emerged. Vertebrates developed into a plethora of backboned fishes, then around 350 million years back, a few descendants of those fishes – pivotal animals in Mammalia's backstory – became the first four-legged creatures capable of forays across dry land.

All that history is fundamental to mammals existing, but – except for Chapter 4's brief consideration of how fish left the water – I will take most of it for granted. Instead, this book will focus on the biological history that is unique to mammals and their pre-mammalian ancestors. It covers, therefore, roughly 310 million years.

That equates to nearly a million years per page or about 800 years per letter. I have no idea how to do such a timeframe justice; I'm not even entirely sure there's a way to truly, humanly understand what such numbers mean. If those 310 million years had been filmed, and the video was then played back sped up three billion times – so that a century passed by in just a second – you'd still have to watch a five-week-long movie.

Three-hundred-and-ten million years ago is when the last ancestor that you and I share with a crocodile lived. Which is to say, it was when the mammalian lineage diverged from the one that led to our closest living relatives, the reptiles. Since this 310-million-year-old split, reptiles have done their thing – evolving into lizards and snakes, turtles, crocodiles, dinosaurs, and their surviving descendants, the birds[*] – and the mammalian

---

[*] Because of this ancestry, a bird is a reptile; the lineage that gained feathers and learnt to fly is completely nested in the reptilian family tree. They have, however, evolved so many characteristics that distinguish them from other reptiles, that I'll maintain the everyday habit of referring to reptiles and birds separately.

lineage has independently gone about its own business, yielding, well, just mammals (see the illustration on the inside front cover).

This history can most profitably be split into three chunks. The first is the pre-mammalian phase that ran from 310 to 210 million years ago. This 100-million-year stretch of evolution began with that last ancestor we mammals shared with crocodiles – a lizard-like animal that looked decidedly more reptilian than it did mammalian – and ended with an animal that was unmistakably a member of our clan.

Early travellers on the mammalian fork were not mammal-like at all, their fossils distinguished from those of reptilian ancestors only by a certain hole in their skulls. Perhaps the most famous of these was *Dimetrodon*. You've likely seen images or models of *Dimetrodon* – it looked like a large lizard with a giant sail on its back. As a kid, I had a rubbery orange *Dimetrodon* that I kept in a box with my toy dinosaurs. I had no idea I was perpetuating an infuriatingly common mistake – no one should confuse a mammalian ancestor with a dinosaur. Not only were dimetrodons on a different branch of the vertebrate family tree, they also lived scores of millions of years before dinosaurs existed, during a time when mammals' ancestors dominated reptilian forebears.

To say the first mammal lived 210 million years ago is to go by the trait most popularly used to distinguish true mammals from their pre-mammalian ancestors: a lower jaw composed of a single bone that forms a particular joint with the skull. If this seems arbitrary, it is, on the one hand, a commonly fossilised feature that came into existence at a time when the mammalian lineage had developed most of the other skeletal characteristics that today define mammals. And on the other, it is an indicator that animals were living increasingly high-energy, warm-blooded lifestyles benefiting from ever more efficient jaws.

Strictly, once the first mammal lived, the process of evolving mammals was finished – the elemental traits that define mammals were in place, to be inherited by each animal's descendants and to ensure all of today's mammals' alikeness. And, so, this initial 100-million-year period is the crux of this book. All experiments in mammalian-ness since then have been nature jamming on a riff.

The second phase runs from 210 million years ago through to 66 million years ago. The first mammals evolved not long after the first dinosaurs emerged, meaning that for the initial two-thirds of mammals' existence, they lived alongside – or beneath, above, or otherwise avoiding – these domineering reptilian neighbours. Living like this had some pretty severe consequences for ecological opportunity. While dinosaurs were around, no mammal was ever bigger than a badger. However, fossils found in the last two decades have shown that our pocket-sized ancestors were much more diverse than people had ever previously expected.

The alternative to using a jaw joint to classify an animal as a mammal is to say that a mammal is any animal descended from the last common ancestor of all of today's mammals. No one can agree on exactly when the last ancestor shared by monotremes, marsupials and placental mammals lived; estimates range from around 161 to as far back as 217 million years ago. For convenience's sake, I'll from now on use a recent estimate that monotremes branched off from the line leading to both marsupials and placental mammals around 166 million years ago.

As we'll discuss in Chapter 2, biologists often look to monotremes – the platypus and the echidnas – to help infer what early mammals were like. The five living monotreme species are the only mammals that currently lay eggs, as the first mammals and all their ancestors did. Live birth evolved 50 million years or so after the origin of mammals, but as it's now how all but these five mammalian species so distinctively reproduce, it is a focus of Chapters 5 and 6 of this book.

From today's perspective, the non-monotreme mammalian lineage subsequently splitting into marsupials and placental mammals is the most notable divergence. These two tribes separated perhaps 148 million years ago.*

Today, there are 19 different types – or orders – of placental mammals: categories such as rodents, bats, carnivores (including cats, dogs and bears), primates, elephants, armadillos and manatees – see the figure on the inside back cover. Since the adoption of evolutionary classification in the mid-nineteenth century through to the 1990s, the relationship between these groups was inferred from their physical likenesses, but as we'll discuss in Chapter 9, in the 1990s, DNA analysis completely redrew the mammalian family tree. The new genealogy was shocking, but so far, nobody has been able to refute it. And the new family tree rather elegantly conjoins genetics with geography, serving as a reminder that evolutionary timescales are on a par with plate tectonics.

Phase Two of mammalian history ended when a fateful meteor struck the earth in the Gulf of Mexico 66 million years ago to end the dinosaurs' reign. The demise of these animals left the world suddenly much more open to mammals living and evolving. And they took full advantage. No matter how varied we now know mammalian contemporaries of dinosaurs to have been, once those giant lizards were obliterated, mammalian diversity exploded. In rocks that sit directly above the layer of meteor dust marking the dinosaurs' extinction, there lies an orgy of mammalian creativity. If nature jams on individual riffs, it was as if some competent but staid amateur musicians handed over their instruments to John Coltrane and Thelonious Monk. Boom: the age of mammals began.

---

* Again, estimates of the date marsupial and placental mammals separated vary considerably (from around 143 to 178 million years past), as they do for nearly all major events buried in the deep, deep biological past. Throughout the book, I have, though, used the most popular estimates without discussing the range of possibilities around them.

CHAPTER ONE

The Descent of Man('s Gonads)

Football fans call it 'brave goalkeeping', the act of springing into a star shape in front of an attacker who is about to kick the ball as hard as possible towards the goal. As I shuffled from the field, bent forward, eyes watering, waiting for the excruciating whack of pain in my crotch to metamorphose into a gut-wrenching ache, I thought only *stupid goalkeeping*. But after the fourth customary slap on the back from a teammate chortling, 'Hope you never wanted kids, pal,' my only thought was *stupid, stupid testicles.*

Natural selection has sculpted the mammalian forelimb into horses' front legs, dolphins' fins, bats' wings, and my football-catching hands. Why, along the mammalian lineage, did evolution house the essential male reproductive organs in a delicate exposed sac? It's like a bank deciding against a vault and keeping its money in a tent on the pavement.

Some of you may be thinking that there is a simple answer: temperature. This arrangement evolved to keep testes cool. I thought so too and assumed that a quick glimpse at the scientific literature would reveal the biological reasons for sperm's sensitivity to warmth and that I'd quickly move on. But what I found was that the small band of scientists who have dedicated their professional time to pondering the scrotum's existence are starkly divided over this so-called 'cooling hypothesis'.

Reams of data show that scrotal sperm factories, including humans', work best a few degrees below core body temperature. And the scrotum is today a finely honed contraption of ligaments that can lift and lower their cargo and contractible skin, which together function to keep testicular temperature a little lower than that of the core body. (In humans that difference is -2.7°C, or nearly -5°F.) The problem is, this doesn't prove cooling was the reason that testicles originally descended. It's a chicken-and-egg situation – did testicles leave the kitchen because they couldn't stand the heat, or do they work best in the cold because they had to leave the body?

All my other vital organs work optimally at 37°C, and most of them get bony protection: my brain and heart are shielded by skull and ribs, while my wife's pelvis protects her ovaries. Forgoing skeletal protection is dangerous. Because male gonads are suspended chandelier-like on a flexible twine of tubes and cords, every year thousands of men go to hospital with testicular ruptures and torsions. But having exposed testicles as an adult is not even the most dangerous aspect of our reproductive organs' arrangement.

The developmental journey to the scrotum is treacherous. At eight weeks of development, a human foetus has two unisex structures that will become either testicles or ovaries. In girls, they don't stray far from this starting point up by the kidneys. But in boys, the nascent gonads make a seven-week voyage across the abdomen on a pulley system of muscles and ligaments. They then sit for a few weeks before coordinated waves of muscular contractions force them out through the inguinal canal.

The complexity of this journey means that it frequently goes wrong. About three per cent of male infants are born with undescended testicles, and although often this eventually self-corrects, it persists in around one per cent of one-year-old boys and typically leads to infertility.

Excavating the inguinal canal also introduces a significant weakness in the abdominal wall, a passage through which internal organs can slip. In the US, more than 600,000 surgeries are performed annually to repair inguinal hernias – the vast majority of them in the scrotal sex.

This increased risk of hernias and sterilising mishaps seems hardly in keeping with the idea of evolution as 'survival of the fittest'. Natural selection's tagline reflects the importance of attributes that help keep species alive – not dying being an essential part of reproductive success. How can a trait such as scrotality (to use the scientific term for possessing a scrotum), with all the handicaps it confers, fit into this framework? Its story is certainly going to be less straightforward than the evolution of a cheetah's leg muscles. Most investigators have tended to think that the advantages of this curious anatomical arrangement must come in the shape of improved fertility. But this is far from proven.

Beyond the big why-have-a-scrotum question, sperm and testicle biology bursts with quite sensible examples of adaptations to animals' lifestyles. Take, for instance, males that contend with sperm competition. In such species – which are numerous among mammals – females mate with a number of males and fatherhood is determined by which sperm win the swimming race. (Not to say that there is always a single champ – in shrews, for example, pups in a single brood often have different fathers.) Chimpanzees mate like this, whereas gorillas have a system in which the dominant silverback gets sole access to the females. The results? First, chimpanzees have sperm that swim considerably faster than their gorilla counterparts and, second, chimp testicles are four times larger than the diminutive ones of gorillas.

Chimps and gorillas, however, came along only about 140 million years after the first scrotum. When considering any evolved characteristic, good first questions to ask include who has it now and, crucially, who had it first. In the case of scrotums, the answer to this latter question must, of course, be inferred, because these fleshy structures do not fossilise. Everything must be surmised from a survey of today's diversity and what is known of these animals' history – plus increasingly from a peak at the genetics underlying this curious trait.

The scrotum has nothing to do with the origins of testicles. Sex has been a fixture of life for a long, long time, pre-dating the split between animals and plants. In all animals, be they baboon or blue tit, cod or crocodile, frog or fruit fly, the male of the species has a pair of testes for producing his seed.* But in birds, reptiles, fish, amphibians and insects the male gonads

---

* The world's largest testes by body weight belong to the tuberous bush cricket, where they amount to 14 per cent of the male's mass; in humans they are around 0.06 per cent.

are internal – where you'd expect these indispensable organs to be.

The scrotum is a curiosity unique to mammals. What's needed, therefore, is a ball's eye view of the mammalian family tree. Thankfully, in 2010, a team from Prague took the most up-to-date reconstruction of mammalian heritage and pasted on it – like fruit hanging from a tree – data from anatomists who have examined the bathing-suit area. What this revealed was that the monumental testicular descent occurred pretty early in mammalian evolution. And what's more, the scrotum was so important that it evolved not once, but twice.

We know that the first mammals lived about 210 million years ago and that the egg-laying platypus and echidnas branched off the mammalian mainline about 166 million years ago. As a result, apart from the eggs, these animals have many key mammalian features such as warm blood, fur and lactation, but each of them is a little 'off'. For example, platypuses' and echidnas' average body temperatures are rather low, and they essentially sweat milk rather than having tidy nipples. I'll include plenty more on this later. What's important now is that platypus and echidna testicles sit – just like those of all early mammals almost certainly sat – right where they start life, safely tucked in by the kidneys.

Then about 20 million years after the ancestors of platypuses and echidnas had gone their way, mammals split again into their two main camps – placental types and marsupial types. And it is on the marsupial branch of the family tree that we find the first owner of a scrotum. What that animal's parents made of it, we will never know.

Nearly all marsupials today have scrotums, and so, logically, the common ancestor of kangaroos, koalas and Tasmanian devils had the first. Marsupials evolved their scrotum independently from us placental mammals, something we know thanks to a host of technical details, the most convincing of which is that it's back to front. Marsupials' testicles hang *in front* of their penises.

THE DESCENT OF MAN('S GONADS)   31

About 50 million years after the marsupial split comes –
scrotally speaking – the most interesting fork in the placental
mammal family tree. Take a left turn, and you will encounter
elephants, mammoths, aardvarks, manatees, hyraxes, and
groups of African shrew-like, hedgehog-like and mole-like
creatures. But you will never see a scrotum – all of these
animals, like platypuses, retain their gonads close to their
kidneys (except, curiously, aardvarks whose testicles lie low
down in their abdomens). And you will also never find a
scrotum among the South American sloths, anteaters and
armadillos, which branched off early too.

Take a right turn, however, to the human side of the tree
at this 100-million-year-old juncture and you'll find external
testicles everywhere. Whatever they're for, scrotums bounce
along between the hind limbs of cats, dogs, horses, bears,
camels, sheep and pigs. And, of course, we and all our primate
brethren have them. This means that at the base of this branch
is the second mammal to independently concoct scrotality –
the one to whom we owe thanks for our dangling parts being,
surely correctly, behind the penis.*

Within the finer branches of this part of the mammalian
family tree is, however, where it gets interesting, for there are
numerous groups – our descended but ascrotal cousins – whose
testes drop down away from the kidneys but don't exit the
abdomen. Almost certainly, these animals evolved from ancestors
whose testes were external, which means at some point they
backtracked on scrotality, evolving anew gonads inside the
abdomen. They are a ragtag bunch, including hedgehogs, moles,
rhinos and tapirs, hippopotamuses, dolphins and whales, some
seals and walruses, and pangolins, the scaly anteaters.

For mammals that returned to the water, tucking everything back up inside seems only sensible; a dangling scrotum

---

* Strangely, the scrotums of rabbits and hares lie, like those of
marsupials, in front of their penises. A rather curious anatomical trait
that has been used to argue that these animals might actually be
closely related to marsupials. (They're not.) And while we're here, the
scrotum of the yellow-bellied house bat is reportedly behind its anus.

isn't hydrodynamic, and it would surely be an easy snack for fish attacking from below. I say snack, but the world record-holders, right whales, have testicles that tip the scales at more than half a tonne each.* The trickier question, which may well be essential for understanding the scrotal sac's function, is: why did it lose its magic for terrestrial hedgehogs, rhinos and scaly anteaters?

Cool idea

The scientific search to explain the scrotum's *raison d'être* began in England in the 1890s at the University of Cambridge. Joseph Griffiths, using terriers as his unfortunate subjects, pushed their testicles back into their abdomens and sutured them there. As little as a week later, he found that the testes had degenerated, the tubules where sperm production occurs had constricted, and sperm were virtually absent. He put this down to the higher temperature of the abdomen, and the cooling hypothesis was born.

The first issue that this hypothesis faced was that it might not have been the abdomen's warmth that was the problem; perhaps something else, a chemical found there, say, had caused the tissue damage. This issue was neatly resolved in the 1920s, during something of a golden age of testicular research. A Dr Fukui, in Japan, repeated Griffiths' experiment but applied a small cooling device to the abdominal wall above the sutures and showed that this prevented the degeneration.

Also in the 1920s, research spearheaded by Carl Moore in Chicago used techniques from the quickly maturing field of cell biology to outline the primary ways in which elevated temperatures interfered with sperm production. Armed with

---

* Blue whales – which weigh two to three times as much as right whales – normally have the biggest everything, but their testicles weigh about ten times less than those of right whales. The preposterous enormity of right whale gonads is probably related to their highly promiscuous lifestyles.

these findings at a time when Darwin's ideas about natural selection were sweeping through biology, Moore was the first to frame the cooling hypothesis in frank evolutionary terms. He argued that after mammals had transitioned from cold- to warm-blooded, maintaining their bodies at constantly elevated temperatures would have severely hampered sperm production, and the first mammals to cool things off with a scrotum became the more successful breeders.

Heat disrupts sperm production so effectively that biology textbooks and medical tracts alike give cooling as the reason for the scrotum. The problem is, many biologists who think seriously about animal evolution are unhappy with this. Opponents say that testicles function optimally at cooler temperatures because they evolved this trait *after* their exile.

If mammals became warm-blooded 210 million or more years ago, it would mean placental mammals carried their gonads internally for over 100 million years before the scrotum made its entrance. The two events were hardly tightly coupled.

The biggest problem for the cooling hypothesis, though, is all the sac-less branches on the family tree. Regardless of their testicular arrangements, all mammals have elevated core temperatures. If numerous mammals lack a scrotum, there is nothing fundamentally incompatible with making sperm at high temperatures. Ascrotal elephants, for example, have a higher core temperature than gorillas and most marsupials. And beyond mammals it gets worse: birds, the only other warm-blooded animals, have internal testes despite having core temperatures that in some species run to 42°C. If cooling is so important, how can so many animals get by with internal testes? The coolers' only riposte to the bird issue is to say we can't learn anything from birds because they're just too different, too evolutionarily distant, from mammals – or possibly that birds' internal air sacs help cool their testes. Things got interesting when it was discovered that dolphins, and some seals, seem to have a special internal cooling system for their reabsorbed gonads; incoming blood is freshened by the testicular arteries intermingling with abdominal veins that bring cooled blood back from the tail and dorsal fin. But seals and dolphins evolved from scrotal ancestors whose testicles would have been adapted to cooler temperatures.

For years, biologists have warmed scrotal testicles and seen them work less well, but no one, it seems, has taken an elephant testicle (or, perhaps, a more manageable African golden mole one) that resides near the kidney and shown that it works better at a cooler temperature. Chances are it wouldn't. Many of the proteins that operate in testicles are also needed by other cell types throughout the body. Normally, all tissues, liver, kidney or leg, use the same gene to make that protein. But protein function is very temperature-dependent, and studies mapping the genes for proteins put to work by the testicles have found that, for a number of them, the genome contains two forms: the body's one, which makes a protein that functions optimally at 37°C, and a modified gene, which builds a protein specialised for work in cooler climes in the scrotum. This strongly suggests that early scrotums had to rely on blunt tools – proteins designed to work at core temperature. Evolution's gradual creation of testes-specific proteins is evidence that externalised testicles had to change to adapt to a cooler life.

However, a recent analysis of testicle position and *precise* body temperature across mammals may amount to the strongest support yet for the cooling hypothesis. Barry Lovegrove in Durban suggested in 2014 that a final spike in core body temperature that occurred as mammal evolution took off after the dinosaurs departed may have necessitated the evolution of the scrotum. In this scenario, mammals were warm-blooded for 150 million years but at a slightly cooler temperature – around 34°C – and scrotal testicles continued to work at that original temperature, leaving the body only after its temperature increased further. Not all the data line up (they never quite seem to) but a lot do – *most* mammals without descended testicles are just a fraction cooler than *most* scrotal ones.

But the other big problem with the cooling hypothesis is that the scrotum is a complex unit built by a multifaceted developmental process, and the evolution of such things does not happen suddenly. A manatee will never suddenly give birth to a son with a scrotum; rather scrotality occurs incrementally. Opponents of Darwin often argued, how can evolution build an eye? What use is half an eye? To counter such arguments, biologists must – as Darwin attempted to – account for why all intermediate stages were desirable.

Today, with improved knowledge of how eyes work, arguments can be convincingly made for how a light-sensitive patch of skin incrementally evolved into the wondrous devices we now have in the fronts of our heads, with all intermediate types having been useful to whoever possessed them. For testicular descent into a scrotum, a similar schema is required. At the very least, we know that the ancestor of scrotal mammals must have had descended but ascrotal testicles and if, as is the case with living animals of this type, there is no cooling in such an arrangement, why was this advantageous to its possessor? Cooling cannot account for an initial move away from the kidneys. It's not to say that one thing couldn't have driven this and then the need for chilling made a scrotum, but one factor driving both steps would, perhaps, be more pleasing.

On top of this, an argument for why cooling would be better for sperm has to say exactly why. The idea that a little less heat might keep sperm DNA from mutating has been proposed, and recently it's been suggested that keeping sperm cool may allow the warmth of a vagina to act as an extra activating signal. But these ideas still fail to surmount the main objections to the cooling hypothesis.

Finally, Michael Bedford of Cornell Medical College, US, is no fan of the cooling hypothesis applied to testicles, but he does wonder whether having a cooled epididymis, the tube where sperm sit after leaving their testicular birthplace, might be important. (Sperm are impotent on exiting the testes and need a few final modifications while in the epididymis.) Bedford has noted that some animals with abdominal testes have extended their epididymis to just below the skin, and that certain animals' furry scrotums have a bald patch for heat loss directly above this storage tube. But if having a cool epididymis is the main goal, why throw the testicles out with it?

Seeking alternatives

If the scrotum's purpose is not to cool organs essential for propagating mammalian life, what is its purpose? Wading into the literature on this subject, you might not find anything with the intuitive appeal of the battered cooling hypothesis, and perhaps nothing here is problem-free either, but there are a couple of intriguing possibilities.

One alternative to scrotums benefiting sperm is that, despite their fragility, they actually benefit their owner. Such a notion was first presented in 1952 by the Swiss zoologist Adolf Portmann after he'd presented the first major attack on the cooling hypothesis. What he proposed instead was the 'display hypothesis'. Portmann argued that by placing the gonads on the outside, the male was giving a clear indication of his 'reproductive pole', a sexual signal important in inter-gender communication. Portmann's best evidence was a few Old World monkeys who have brightly coloured scrotums.

This theory is not widely accepted because such conspicuous displays are rare (many scrotums are barely visible) and bright coloration evolved long after the original scrotum. Some say it's not surprising that in its 100-million-year existence, the scrotum has been co-opted as a sexual attractant by a handful of groups.

I was just about to disregard completely the display hypothesis when two things happened. First, a colleague returned from her honeymoon in Tanzania excitedly showing to anyone who'd look photos of a scrotum. The scrotum belonged – don't worry – to one of Portmann's Old World monkeys, a vervet monkey, and it was screamingly, beguilingly bright blue.

OK, it's just one monkey, I thought, but then I met Richard Dawkins. I had three minutes with the esteemed evolutionary biologist at a book signing, so I asked him for his opinions on the scrotum. After expressing serious doubts about the cooling hypothesis, he said he wondered whether it might have something to do with evolutionary biology's handicap principle.

Handicap theory posits that if a female had to choose between two suitors who had beaten off all other competitors, but one had done so with a hand tied behind his back, she'd go for him because he's obviously tougher still. It is controversial, but it does offer explanations for a number of problematic biological phenomena, such as male birds' colourful plumage and songs that should attract predators. If the handicap theory is right, the scrotum exists to let its possessor say, 'I'm *so* able to look after myself, I can keep these on the outside!'

Regarding scrotums, this theory doesn't have a whole lot of backers, although the idea is not dead. A recent study of prairie voles, for example, found that female voles do, in fact, prefer males with larger testicles, and this is a species where the scrotum is very much undecorated.

Curiously, however, the original proponent of the handicap theory in general – an Israeli biologist named Amotz Zahavi – didn't himself favour it for the scrotum. He instead came up with the 'training hypothesis', although it was only Zahavi's off-the-record sharing of this idea with a colleague named Scott Freeman that resulted in it entering the public domain, with Freeman writing it up for the *Journal of Theoretical Biology* in 1990.

This idea has it that the scrotum, with its poor blood supply, keeps the testicles in an oxygen-starved environment, which toughens up the sperm. Deprived of this essential gas, sperm react in various ways that make them better prepared for the Herculean task of ascending a vagina, a cervix, a uterus and a fallopian tube.

Freeman went to great lengths to look at the size of numerous species' crown jewels and found that there is an excellent correlation between testicular size and the number of sperm in an ejaculate, and more surprisingly that, overall, internal testicles are larger than descended ones. The take-home message was that the 'training' produced a quality–quantity trade-off – scrotal animals could make fewer sperm because they made better sperm.

Freeman must be credited for revealing these interesting correlations, but the problem with the training hypothesis is that it's mainly concerned with the testicles' lousy blood supply rather than their expulsion. One can't help but think that it would have been easier to evolve poor gonadal vasculature while keeping them in the body.

A few years later, in the mid-1990s, Michael Chance, a professor of animal behaviour at the UK University of Birmingham, came across a newspaper story about the Oxford–Cambridge University boat race that piqued his interest in testicles. He learnt that, after the race, the rowers' urine contained fluid from their prostates.

The oarsmen's exertions – their cyclical abdominal straining – had deposited prostatic fluid in their urethras because there are no sphincters in the reproductive tract. Without such valves of circular muscles, squeezing any of the sacs and tubes that make up this system is liable to rearrange its contents. In 1996, in what has become known as the 'galloping hypothesis', Chance argued that externalisation of the testes was necessary when mammals started to move in ways that sharply increased abdominal pressure.

A survey of how mammals move reveals a good deal of variety. And when Chance listed animals with internal testicles, he didn't find many gallopers. The elephants, aardvarks, and their cousins on the undescended branch of the mammalian tree don't bound or jump around. On the other side, the creatures such as moles and hedgehogs that reabsorbed their sexual cargo seem to have evolved away from internally disruptive types of movement. Among mammals that have returned to the sea, the few that have retained scrotums are the only ones who breed on land, such as elephant seals, who fight vigorously to defend their territory during the rutting season.

One might argue that evolution could surely have thrown in a sphincter or two, or some internal shielding, but besides the possibility that the mechanics of ejaculation would struggle with such things, another argument supports Chance's thinking. In 1991 Roland Frey of Germany's Freiburg University wrote a paper (which Chance had apparently not read) that also argued that testicle externalisation had been driven by rising intrabdominal pressure. In it, he described a number of features of blood vessels of scrotal testes that ensure more constant pressure, possibly to avoid impaired blood drainage during galloping. The specific adaptations are different between marsupials and the rest of us, but seem aimed at the same goal.

The galloping hypothesis would be a case of evolutionary compromise – the dangers of scrotality being a necessary price for the greater advantages of a new and valuable type of movement. Also, if a move away from the flexing and extending vertebral column had provided some degree of pressure buffering, then this idea alone has the appeal of offering plausible advantages for a descended but ascrotal arrangement.

There are many theories in evolutionary biology. Often there's great pleasure in the detective-like process of piecing together the available, incomplete evidence into a coherent story, but the big challenge for this science is actually testing ideas. One exciting recent development that is providing new data on the behaviour of mammals' nomadic testicles has been the identification of the signal that controls the testicles' initial descent from the kidney region to the undercarriage.

When the testes and ovaries are young, they are held in place by the so-called cranial suspensory ligament, while holding on loosely is a second, measly ligament termed the gubernaculum. To begin their roller-coaster ride, testicles secrete a signal that causes the suspensory ligament to degenerate and the gubernaculum to become capable of guiding them to the base of the abdomen.

Remarkably, two groups – one in Germany, the other in Texas in the US – simultaneously uncovered the identity of the testicles' 'come-and-get-me' signal. It's a molecule related to insulin, called (not terribly imaginatively) insulin-like hormone 3, or INSL3. When these scientists deleted the gene for this signal the testes stayed, ovary-like, by the kidneys.

A somewhat ghoulish follow-up experiment asked if ovaries stay put because they do not switch on the INSL3 gene. A handful of female mice were genetically engineered to have high

INSL3 levels in their gonads, and this was enough to get their ovaries pulled down to the foot of their abdomens.

Fascinated by INSL3's role in testicular descent, and a related gene's role in the mammalian speciality of lactation, Teddy Hsu and his colleagues at Stanford University in California turned to the duck-billed platypus. In 2008, they found that the platypus has a single gene for the prototype version of the signal and that it was this gene's duplication in subsequent mammals that allowed one version to evolve a function in testicular descent and the other in nipple development. It's a beautiful example of a genetic event in biological history that helped produce a mammalian specialisation.

What's more, grasping the role of INSL3 and also the cellular receptor that this chemical messenger activates – which goes by the name of RXFP2 – gave evolutionary geneticists two genes that they could examine for clues as to how and when descended testicles first evolved. The results of this were rather surprising.

Michael Hiller and colleagues at the Max Planck Institute for Molecular Cell Biology and Genetics in Dresden, Germany, published in 2018 a survey of the INSL3 and RXFP2 genes from over 70 different species of placental mammals. They detected the genes in all of them but when they looked at the animals on what I called the scrotum-less left turn – so elephants, aardvarks, manatees, hyraxes, African shrew-like, hedgehog-like and mole-like creatures – things got rather interesting. The aardvark, which has descended but ascrotal testes, had functional copies of both genes, but in four of the animals whose testes reside by their kidneys, Hiller's team found broken genes. There were stretches of DNA that were recognisably INSL3 and RXFP2 but these DNA sequences had clearly undergone a series of mutations that had rendered them defunct. These animals had once had functional INSL3 signalling but not anymore; strongly suggesting that they had once also had descended testicles, but no longer.

Rather than these African mammals retaining the original platypus-like testicular state throughout their evolutionary history, they actually – for some reason – re-evolved it via a loss of these genes. This interpretation was strengthened by the observation that the mutations in these genes found in manatees, cape elephant shrews, cape golden moles and lesser hedgehog tenrecs were all decidedly different, indicating that each of these lineages had independently called time on the gubernaculum grabbing their testes.

Hiller believes these findings indicate that the very first placental mammals had descended testicles – at least, getting as far the underside of the abdomen (so not necessarily to a cooler place). One question mark hanging over this conclusion is the fact the South American armadillo and sloth, which have the only slightly descended testes, had intact genes for INSL3 and RXFP2. And as we'll see later in the book, the issue of exactly how the first main branches of the placental mammal family tree split is a matter of debate still. Nevertheless, this study is a beautiful example of how evolutionary changes can leave tell-tale traces not in an animal's anatomy but in its DNA.

A crucial next step will be determining the genes required for forming the inguinal canal and making the scrotum. Probably the best place to look will be in those mammals that have backtracked on externalisation, where these genes have likely been deactivated.

It's rather humbling to realise that this basic aspect of our bodies remains a mystery. The fact that such a ridiculous appendage evolved twice surely means we should be able to get a handle on it. As more findings are accumulated, it would take a brave person to bet against an all-conquering theory of testicle externalisation one day dropping out of all this research – the 'scrotality totality'?

A successful theory will have to explain the full diversity of mammalian testicle positions, not just the scrotum's existence. I like Chance and Frey's galloping hypothesis, but could a scrotum really be the only way to deal with undulating abdominal pressure? And Lovegrove's recent survey does support a role for temperature sensitivity. Signalling is still an outside bet, but if scrotums were actually sexually selected, where's the mammalian equivalent of the peacock, some species toting a pair of footballs?

Talking of which, while we wait for the scrotality totality, we goalkeepers should probably look to our cricket- and baseball-playing friends, who use evolution's gift of a large brain and opposable thumbs to don a protective box.

CHAPTER TWO

Life on the Edge of Mammaldom

Platypuses don't travel well. The one I know best is stuffed and lies in the British Museum. He's male. I know this because he has venom-delivering spurs on his hind legs. The closest a live platypus has come to the UK was in 1943 when Winston Churchill requested one be sent to London in a bid to raise wartime morale. The Australian government dispatched an adult male, but he died in his tank aboard the MV *Port Philip* four days from Liverpool harbour when the ship let off depth charges in response to a submarine attack.

Even the first dead one to reach England is rumoured to have splashed to the ground when the barrel of spirits in which it had travelled burst open atop the head of the woman carrying it from Newcastle docks. This was in 1799 when Australia was a brave new frontier. The barrel had been sent by Captain John Hunter, the second governor of the new Sydney penal colony, and in it there had also been the first wombat ever to leave the country. Hunter had been intent on sending a complete platypus, but after an unseasonably warm spell in New South Wales, his specimen had started to whiff. He had to discard the animal's innards and send just its skin back to the Old World. Hunter tucked in a sketch he'd made of the living animal and a note reading 'small amphibious animal of the mole kind', and with that the long process of making sense of this singular animal began.

For nearly a century, the ensuing debates shaped what exactly defined a mammal. And to this day, platypuses serve as priceless sources of insight into the evolutionary journey that led to the emergence of mammals.

A new species arriving in Europe from a far-flung land in 1799 was nothing unusual. For centuries explorers had sent home curious new beings, but Captain Cook's discovery of Australia's east coast in 1770 produced a transformative set of imports. Londoners loved the first kangaroo. Both naturalists, who wanted new plants and animals to describe, and the public, hungry to be wowed, developed a real appetite for Australia's curiosities.

A tiny-eyed aquatic mole with no external ears, webbed feet, the tail of a beaver and the bill of a duck was, however, too much. Plus, the platypus had arrived via the China Seas where fishermen sewed monkey torsos onto fish tails and peddled their creations as mermaids. Famously, the first academic description of a platypus questioned whether it was a hoax. 'It naturally excites the idea of some deceptive preparation by artificial means,' wrote George Shaw, concluding it was only after 'the most minute and rigid examination that we can persuade ourselves of its being the real beak or snout of a quadruped'.[*]

But when further specimens confirmed the animal's authenticity, the most eminent naturalists of the day clambered to make sense of it. The platypus seemed an outright assault on the neat categories Linnaeus had spent years establishing. Thomas Bewick, who drew it for his popular book *A General History of Quadrupeds*, said that the platypus had 'a threefold nature, that of a fish, a bird, and a quadruped, and is related to nothing we have hitherto seen'.

What really complicated matters was the later arrival of specimens complete with internal organs. When Everard Home, a surgeon and fellow of the Royal Society, published the first detailed descriptions of a male and a female platypus in 1802, he noted that certain characters were entirely mammalian, but that in numerous other respects platypuses resembled birds or reptiles.

Particularly vexing was the female's reproductive anatomy. This was crucial because, when it came to classification, taxonomists cared deeply about how a plant or animal

---

[*] Throughout these decades, 'quadruped' remained the favoured term for a mammal.

reproduced. Whales and dolphins had been welcomed into the mammalian fold, despite their lack of legs and fur, because they birthed live young and lactated. To classify platypuses according to a system defined by plants and animals possessing certain key traits, knowing their mode of birth and whether or not they possessed mammary glands was essential.

This sounds as if it should have been straightforward, but these debates took place 9,500 miles from where platypuses quietly lived in the freshwater habitats of eastern Australia, mating just once a year, prior to the female very privately giving birth closeted deep inside her riverbank burrow. No major figure in these discussions ever saw a live platypus, let alone the birth of one.

Regarding lactation, things appeared clear-cut. Home stated that his female had no nipples and, by implication, no mammary glands; surprising, given the animal's fur, but Home was unambiguous about this. Further down the body, however, Home was truly stumped. The female's reproductive tract was like nothing he'd ever seen. It was impossible to tell if the arrangement of tubes he encountered laid eggs or birthed live youngsters.

So important was reproduction, Home acquired numerous other less-studied animals to try and find a match. Quickly, he saw that the platypus's reproductive anatomy resembled that of echidnas – the spiny anteaters who'd arrived from Australia a decade earlier with none of the fanfare the strange duckbill got. Based on their anatomical kinship, Home said these two species represented a 'new tribe of animals'.

This didn't, though, tell him how the system worked, and so he looked further afield. Eventually, Home declared that his new tribe most closely resembled certain snakes and lizards whose young developed inside eggs, but in eggs that were retained within the mother until they hatched. They were ovoviviparous. That way, platypuses were born, as mammals were, without shells.

Home made many astute observations and drew many learned conclusions about the platypus. But regarding both lactation and birth, he was entirely wrong.

Soon, the prominent French naturalist, Étienne Geoffroy Saint-Hilaire, entered the debate. Geoffroy, as he's known, christened this 'new tribe' according to their unusual reproductive arrangement, naming platypuses and echidnas *Monotremata* or monotremes. *Mono* is Greek for one, while *treme* means hole. These animals were named for the fact that they – like birds and reptiles – urinate, defecate and reproduce through a single rear opening. This was part of Home's confusion.

(I confess that I sniggered when I learnt this and briefly felt rather superior. But then I saw that such snobbery was a bit rich coming from a two-holed mammal. Dedicated plumbing for each of these three needs seems to be another example of greater female sophistication. If I ever own a pub, I'll put *Bitremes* and *Tritremes* on the toilet doors.)

Geoffroy was absolute about two things: he agreed with Home that platypuses did not lactate, but he was certain too that they laid eggs. He stated that only two separate modes of reproduction existed: the mammalian one that employed lactation and live birth and another that involved eggs and zero dairy products. Because platypuses bred by the latter route, it meant for Geoffroy that monotremes were not mammals.

Jean-Baptiste Lamarck also felt that platypuses' lack of mammary glands negated their claim to being mammals. Lamarck is maligned today for his disproved theory of evolution based on the inheritance of acquired characteristics when he should probably be celebrated for having championed evolutionary change as early as he did. By virtue of this viewpoint, Lamarck happily and presciently regarded platypuses as intermediates between reptiles and mammals.

Opposing Geoffroy and Lamarck, other prominent scientists argued that platypuses *were* mammals. For some, monotremes sat on the bottom rung of an ascending scale of mammals, with marsupials directly above them and placental mammals above marsupials. Placentals themselves were arranged with primates at the apex.

The quest to resolve these debates, and to definitively establish how platypuses reproduced, became remarkably bitter and really brought out the worst in people. Ideas that had first been presented tentatively reappeared as statements of fact. For instance, Home asserted in 1819 that monotremes were ovoviviparous with a forthrightness alien to the uncertainty of his original report. More problematically, though, when actual new observations were made, senior figures interpreted them – often outlandishly – in ways that supported their prior convictions.

When the German anatomist Johann Meckel mentioned in an 1824 publication that he'd seen platypus mammary glands, Geoffroy claimed that they were likely to be glands for scent signalling or releasing substances that conditioned the fur.

Then, when Meckel published a full description of the mammaries two years later, Geoffroy reasserted that animals came as 'Mammals, Monotremes, Birds, Reptiles, and Fishes', and Home procured new specimens from Australia to ensure he and Geoffroy were correct. Home had his assistant assure him that there were indeed no mammary glands, then finally accused Meckel of being so convinced that platypuses lactated that he had imagined the glands.

To be fair, these disagreements may have stemmed, at least in part, from the fact that platypus mammary glands grow and shrink enormously according to season and usage. Timing would have been everything.

It's fortunate, therefore, that in 1831 the 39th Regiment of the British Army was stationed in New South Wales during platypus breeding season.

With presumably limited military duties, Lieutenant Lauderdale Maule took a mother and two young platypuses from their nest and attempted to keep them by feeding them worms, bread and milk. But after two weeks an unspecified accident killed the mother, and Maule immediately skinned her. In doing so, he saw milk ooze from her nipple-less abdomen, confirming that platypuses lactated.

Maule straightaway wrote to London to report this observation, and the word of an army lieutenant was good enough for the naturalists back home. After years of speculation and fighting, the platypus's standing as a lactating mammal was secured by an army officer terrible at keeping pets.

After their mammary glands secured monotremes' standing as mammals, attention shifted to whether a mammal could lay an egg. Geoffroy – who greeted Maule's observations by spluttering, 'If that's milk, show me the butter!' – remained convinced until his death in 1844 that platypuses *did* lay eggs. But, unsurprisingly, the discovery of platypus mammary glands led many people to conclude that the animals must, therefore, birth live young.

One such person was the young Englishman Richard Owen. Owen was a fine anatomist who had helped to confirm the reality of monotreme mammary glands. He would go on to coin the term 'dinosaur', found London's Natural History Museum, and attempt to convince Darwin that evolution didn't happen. Owen's was an interesting life. And, in being certain that platypuses *didn't* lay eggs, he assumed with aplomb Geoffroy's role of providing unlikely explanations for others' first-hand observations.

For example, when Lieutenant Maule also claimed to have seen fragments of eggshells in platypus nests, Owen attacked his uncertainty, while another sceptic explained that the fragments had probably been excrement. When a letter arrived from Australia in 1864, stating that a captive pregnant female had laid two eggs, Owen said that this was no natural event and that the female must have aborted her pregnancy due to fear.

Only in 1884 would an 80-year-old Owen begrudgingly change his mind in response to William Caldwell, a young zoologist from Cambridge University, working in Australia.

Caldwell resolved a nearly century-long debate, but it's hard to like him. He achieved this by paying scores of local aboriginals to hunt platypuses and echidnas, a tactic that led to a 'fantastic slaughter'. And as the aboriginals brought him more and more animals, Caldwell raised the price of the food he sold them. 'The half-crowns were, therefore,' he wrote, 'always just enough to buy food to keep the lazy blacks hungry.'

More than 1,400 monotremes had been killed before Caldwell himself shot a female platypus in the process of laying her eggs; the first egg lay beside her corpse, the second was still in her dilated uterus.

Caldwell triumphantly composed a telegram for an upcoming gathering of the British Association in Montreal: 'Monotremes oviparous, ovum meroblastic.' This meant: platypuses lay eggs, and the cells inside the egg divide as they do in birds, not in mammals.

Coincidentally, on the exact same day that Caldwell's telegram was read out in Canada, Wilhelm Haacke, head of the Natural History Museum in Adelaide, showed the Royal Society of South Australia fragments of eggshell he'd found in the pouch of an echidna. And so on 2 September 1884, it was doubly confirmed that certain mammals do lay eggs.

A thin straggling branch

In the midst of this prolonged saga, on 19 January 1836, Charles Darwin woke up and went kangaroo hunting. The *Beagle* was anchored in Sydney Cove, having crossed the Pacific from the Galapagos Islands, and 26-year-old Darwin was at a small farmstead in Wallerawang, New South Wales, having detoured there on a trip to the Blue Mountains. He'd found Sydney interesting but had been glad to leave; his journal describes the Australian landscape with far greater enthusiasm than it does the bustling colonial capital.

Darwin didn't see a single kangaroo that day. But he wrote that he enjoyed the gallop and had inspected a kangaroo rat that the greyhounds had chased into a hollow tree. He also worried that these imported dogs must menace the local animals.

Darwin's journal then records how, early in the evening, he lay on a sunny bank and 'reflected on the strange character of the animals of the country as compared to the rest of the world'. Evolution by natural selection was not, as is often portrayed, an idea that came to Darwin in the Galapagos Islands – in Australia, everything was still in flux.

On that bank, the young naturalist pondered the distinct forms animals took in Australia, yet he noted how they resembled creatures elsewhere. He wondered what creative process might have produced the different forms and famously questioned divine Creation.*

Then, 'in the dusk of the evening', he strolled along 'a chain of ponds' and watched some of the 'famous' local platypuses. Darwin watched them 'playing and diving about the surface of the water', and they reminded him of British water rats. The scene reads quite idyllically until his kangaroo-hunting host shoots him one to examine.

Twenty-three years later, *On the Origin of Species* offered Darwin's answers to the questions he'd posed himself on that sunny bank, and the animals he'd watched in that dusk featured in several places.

The rigid taxonomies that had preceded Darwin – those that had struggled to accommodate the platypus – had helped him greatly by arranging plants and animals according to their likenesses. They'd made it clear that different species shared certain characters and varied by degrees from one another. But Darwin was unsentimental. In presenting his theories, he scythed down the philosophy behind former classifications, writing, 'All true classification is genealogical, that community of descent is the hidden bond which naturalists have been unconsciously seeking, and not some unknown plan of creation.'

Origin's sole illustration emphasised this point. On a fold-out page, there was what we now call a phylogenetic tree: a diagram of a biological family tree showing how species die, survive and diverge over geological time. In the diagram, the species

---

* The first edition of *The Voyage of The Beagle* – which anthologised Darwin's travel journals – contained this questioning passage, but Darwin withdrew it from the second edition.

were hypothetical, but the message was clear: all life is somehow related, and this type of chart should form the basis of all future taxonomy.

Darwin argued that the relationships between species could be inferred by observing what features they did and did not share, common features being shared inheritances from the same ancestors. And, if such a character differed between two species, the lineages leading to those species had diverged; with one or both lineages having morphed in ways that made them distinct from the ancestors. The amount of difference was probably influenced how much time had accrued since the two species' separation, and their subsequent circumstances.

In particular, Darwin argued for the usefulness of certain core traits – attributes that were less prone to change with lifestyle. The relatedness of all mammals, he suggested, might be most apparent in the angles of their jaws, their mode of reproduction and their fur – characteristics that would persist even when a mammal superficially changed by, say, evolving a bird-like beak to sift river beds for food (my example).

If the platypus's resemblance to birds and reptiles had horrified many of the people who first described the animal, it must have thrilled Darwin. He was proposing that living beings evolved incrementally. But this gradual mode of change presented a problem. Darwin needed to explain why species or higher-level groups of organisms were discrete and clearly distinguishable entities, instead of life consisting of a smear of gradated forms. Mammals and reptiles, for example, are very clearly different, with many characters distinguishing them and little in the way of intermediate forms lying between them.

The solution, Darwin argued, lay in evolution leading ultimately to the complete replacement of one form by another, so that over geological time there was a succession of types. All intermediate forms *had* existed, but typically they were extinct.

The obvious place to look for in-between forms was therefore among fossils in sedimentary rocks laid down aeons ago. However, Darwin was sceptical about the completeness of the fossil record, and so he considered alternatives. Among them, he wrote that the platypus and the lungfish were 'some of the most anomalous forms now known in the world' and that they might 'fancifully be called living fossils, [which] will aid us in forming a picture of the ancient forms of life'.

To him, these two animals were fortuitous survivors from antiquity. Lungfish were intermediates between fish and amphibians, and the platypus showed a link between reptiles and mammals. For Darwin, they were perfect; their rarity supported his assertion that most intermediate forms had gone extinct; their actual existence was living proof of intermediate forms. When he argued again for drawing a complete tree of life where the branches traced out species' ancestry, he wrote:

> [W]e here and there see a thin straggling branch springing from a fork low down in a tree, and which by some chance has been favoured and is still alive on its summit, so we occasionally see an animal like the [platypus] or [lungfish], which in some small degree connects by its affinities two large branches of life

It's the nature of the thin, straggling monotreme branch[*] that makes platypuses a stalwart of this book – they'll pop up in nearly every chapter. The good fortune that has allowed monotremes to persist is also the good fortune of anyone interested in mammalian history. It seems that most times you try to figure out the natural history of a mammalian trait, examining it in monotremes provides a clue as to how it evolved.

Marsupials and placental mammals evolved from a common ancestor that lived about 148 million years ago, whereas monotremes – the other five species of living mammals – are

---

[*] And the fossil record supports the notion that it's always been a fairly straggly branch – platypuses and echidnas are not the last survivors of a once huge tribe, but rather an evolutionary offshoot.

the sole surviving members of a lineage that sling-shotted off the mammalian mainline about 20 million years earlier. The platypus and echidnas, therefore, provide evidence of what mammalian-ness looked like 20 million years before it reached the form inherited by all other surviving mammals.

If there's a quibble to be had with Darwin's description of the platypus, it lies with the phrase 'living fossil'. That term implies that a living species hasn't changed for a very long time, and that almost never happens. Darwin's suggestion that 'a few old and intermediate parent-forms hav[e] occasionally transmitted to the present day descendants but little modified' is rarely true – things evolve. Platypuses have swum, dived, burrowed and procreated their own meandering path. They've evolved their own characters. One need only look at how different they are to echidnas – despite their common rear-end anatomy – to see that each living tip of this thin mammalian lineage has adapted to its own circumstances. All mammals are not descended from aquatic moles with duck bills.

Instead, aspects of platypuses and echidnas allow us to infer how mammalian traits got to where they are today. As I don't want to spoil any forthcoming monotreme-based surprises, let's take the eggs and nipple-less mammary glands that preoccupied nineteenth-century naturalists.

Imagine if the only animals in the world were platypuses and rabbits, and we wanted to know if their last common ancestor had laid eggs or birthed live young. There'd be little way of telling. The ancestor might have laid eggs, and rabbits evolved live birth, or the ancestor might have birthed live young, and platypuses evolved eggs. Those two scenarios would be equally likely, because each involves one evolutionary transition (see diagram overleaf, top part).

However, if this hypothetical world also contained turtles and we knew from their myriad dissimilarities that the egg-laying turtle lineage had split off before rabbits and platypuses had diverged, we'd have an extra clue. Then, one scenario would be that the last common ancestor of all three laid eggs and only rabbits evolved live birth.

And the other would be that the ancestor birthed live young, and both turtles and platypuses independently developed eggs. The first scheme involves one evolutionary transition, the second requires two (see bottom section of diagram). So, while the latter case is not impossible, the former is simpler. It's more parsimonious.

This principle of parsimony has been at the heart of evolutionary inference for a long time, and it's served the field well. Basically, the simplest explanation is the preferred one.

Also, you see that the more well-characterised groups you have to compare, the more confident you can be in your inferences. Taking the real world, egg-laying is the norm for reptiles, and it's essentially certain that the last shared ancestor

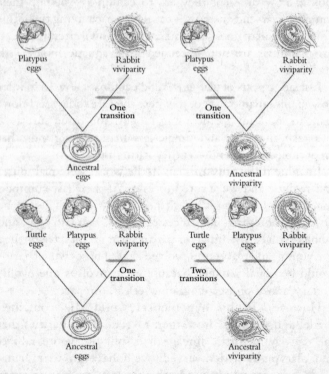

Figure 3.1: Maximum parsimony dictates that the fewest character-state changes makes for the best evolutionary explanation.

of reptiles and mammals laid eggs, with monotremes hugely substantiating this conclusion.

Plus, monotremes' egg-laying gives the evolution of live birth a relative time-stamp. Because of the other features of a platypus, we know that mammals had evolved fur, milk, warm-bloodedness and many other mammalian traits before they started popping out unshelled youngsters.

Turning to mammary glands: no reptile, bird or other vertebrate lactates, so we can be confident that milk evolved specifically in the mammalian lineage. And its presence in placental mammals, marsupials, and – *oui, Monsieur Geoffroy* – monotremes, tells us it had evolved before these three lineages diverged.

This is a conclusion strongly backed by the similarity of mammary gland plumbing, milk composition and the underlying genetics across all mammals. The simplest explanation is that all three lineages inherited their infant-feeding systems from a shared ancestor in which it had already evolved.

But then there's monotremes' lack of nipples. This is the sort of thing I'm talking about when I say these animals provide unique glimpses of a character's evolutionary trajectory. Teatless lactation indicates that, before therians (the collective term for marsupials and placental mammals) came along, mammary glands existed, but in milk's early days the substance leaked from the mother in a more diffuse manner. Young platypuses, for instance, suck at hairs as milk dribbles down them.

Certainly, it makes sense that an animal would not have evolved nipples before it had mammary glands; however, more dispersed milk secretion is potentially an interesting clue as to how and why the mammary gland initially evolved.

Similarly, while monotremes lay eggs, those eggs are intriguingly different from those that birds and reptiles lay. Egg-laying versus live birth tends to get viewed as a stark dichotomy, but monotreme eggs tell us something vital about how exactly mammals transitioned to having squealing babies; a hint at how the gap between eggs and placentas was traversed.

We'll pick up these two threads again in Chapters 6 and Seven.

Defining mammals

It's been said that if mammals have hair and milk, then a coconut is a mammal. Not by anybody important, just someone on an internet forum. I'm not sure whether the remark was intended as an explicit comment on the antiquated and outdated habits of Linnaean thinking, but it was amusing to see how many (strangely earnest) responses rejected the idea that a coconut is a mammal by variously pointing out that coconuts don't have nipples, external ears or three middle ear bones, so, therefore, weren't actually mammals. Additionally, a couple of people, like time travellers from the mid-nineteenth century, also protested that coconuts didn't birth live young. Nobody, however, posted that it was a coconut's genealogical relationship with palm trees, relative to giraffes and warthogs, that disqualified it from being a mammal. In the twenty-first century, Linnaean thinking is alive and well.

If you want more substantive support for this than some online wittering, consult a major dictionary, and you'll find that a mammal is still defined as being an animal with hair, mammary glands and perhaps – depending on the calibre of your dictionary – another trait or two. There's just something about this mode of thinking that chimes with our sensibilities. So much so, of course, that when this particular mammal decided he'd like to write a book about his being such a creature, he decided to build the whole thing on a quasi-Linnaean scaffolding, a fact that right now – knowing that as soon as this chapter's done, I'm straight back to a one-by-one list of definitional mammalian traits – is making me feel uncomfortable. In my defence, I can only suggest that such a conceit allows for attention to be focused on one thing at a time.

An obvious problem with mixing Darwin and Linnaeus – trying to overlay Linnaeus's trait-based scheme on modern biology's genealogically based taxonomy – is that no evolutionary lineage is obliged to maintain a trait that it's inherited. The fact that whale ancestors had four legs and belonged to a group of animals we call tetrapods – exactly for this trait of having four feet – didn't prohibit them from evolving shorter and shorter legs until at some point the back ones disappeared and the front ones became flippers. In biology, if ecological opportunity favours swimming over walking, no one is obliged to stay four-legged out of loyalty to some Linnaean ideal – evolving lineages acquire new traits and shed old ones according to their circumstances.

In fact, dolphins even push the validity of defining mammals by one of their most distinctive features, possessing hair. Elephants, armadillos and humans might all show that having hair doesn't mean you need to have a full-body fur coat, but dolphins take it even further. Dolphins necessitate the caveat 'mammals have *some* hair *for some* of their lives', for dolphins have a few whiskers near their mouths for only the first couple of weeks of their lives (which are thought to aid newborns in finding their mother's nipples).

The other problem with defining types of animals by specific traits is that there's also nothing stopping two separate lineages evolving the same thing independently. For instance, becoming warm-blooded was critical to mammals evolving into the animals that they are, but later a different type of warm-blooded creature also evolved – this one feathered – and warm-bloodedness was no longer strictly a mammalian trait. Sure, a Linnaean approach can emphasise the difference between mammalian and avian modes of insulation, but becoming warm-blooded is the real biological story. And the similarities between mammals and birds don't stop there: the two lines also convergently concocted intensive parenting, large brains and four-chambered hearts. Not to mention that Mammalia also spawned flying bats, and that platypus ancestors grew bills. Convergent evolution can produce remarkable similarities – within mammals you can, for example, look at the many different long-snouted anteaters that have independently emerged – and these confuse matters terribly, not just when people try to construct classifications based on physical traits but also when tracing evolutionary relationships using phylogenetics.

Instead of character-defined groups, Darwinian taxonomists talk about clades. These are groups made up of all the species descended from a common ancestor. This is why, for example, birds are reptiles. Birds evolved from dinosaurs and dinosaurs clearly evolved from the reptilian stock. No matter how many features birds developed to distinguish them from other reptiles, their ancestry remains fixed. More counterintuitively, clade-based classification means that amphibians, reptiles and mammals are all eccentric bony fish. Again, one ancestor gave rise to all bony fish clearly recognisable as such, and mapping its descendants shows that one quirky lineage made a break for the shores and eventually gave rise to all land vertebrates.

Mammals are interesting from a cladistic viewpoint, in that mammalian ancestors diverged from reptilian ancestors 310 million years ago, yet all living mammals derive from a single shared ancestor that lived only about 166 million years ago (see inside back cover). From the origins of the uniquely mammalian lineage to that 166-million-year-old split between monotremes and therian mammals, there are no other surviving side branches from this evolutionary experiment.

That's a big gap, and understanding what happened says something interesting about the nature of pre-mammalian history. The fossil record indicates that there were successive 'radiations' of what have been called different grades of mammalian ancestors, each grade with its own somewhat esoteric name. Opposite is an illustration featuring first, pelycosaurs – of which the sail-backed *Dimetrodon* was one – second, therapsids, who made some major advances in mammalian biology, and third, cynodonts.

By radiation we mean the emergence of a full spectrum of creatures that nevertheless shared a range of traits. Hence,

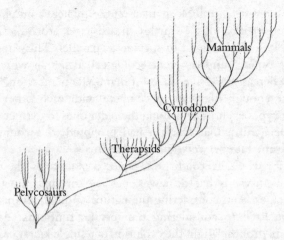

Figure 3.2: Mammals and their ancestors radiating over time.

Dimetrodon was the top-ranked predator among many carnivores in the first pelycosaur radiation, but there were also plant- and insect-eating pelycosaurs of many shapes and sizes. Moving forward through geological time, it seems at least plausible that evolution could have proceeded by herbivores giving rise to future plant-eaters, and carnivores yielding tomorrow's killers, and so on; that is, all lineages meandering side by side into the future. But this isn't what happened. Instead, a single new type of mammalian ancestor emerged from among the pelycosaurs, and from that single lineage nature created a new radiation of meat-, plant- and insect-eating therapsids, which then eventually replaced all the corresponding pelycosaurs.

Then, from the therapsids, another lineage emerged to seed the cynodont radiation, and from there, a line containing the last common ancestor of all mammals was born.*

---

* This pelycosaur–therapsid–cynodont–mammal 1-2-3-4 is a simplification; there were other smaller and overlapping radiations within these major radiations, just as there are today, most obviously marsupials and placental mammals representing two concurrent radiations.

It's interesting to look again at reptiles, where a number of early diversifications — turtles, lizards and crocodiles, for example — all managed to survive in parallel. This suggests reptile types found different ecological niches in which to live, whereas successive waves of mammalian ancestors lived similar enough lifestyles to compete with each other, the newer types eventually driving the older ones to extinction — the mechanism that Darwin had proposed to account for significant gaps between groups of animals.

As for why newer radiations replaced older ones, the most obvious answer is that the newer ones held certain competitive advantages. And considering the nature of these advantages, we are back to considering the specific traits that define different groups. Might the evolution of a single key trait have given therapsids, say, an edge over pelycosaurs? Did a uniquely mammalian character allow mammals to succeed cynodonts?

The answer is a sort of 'yes and no'. Yes, evolving new traits — the subject of this book — yielded certain advantages for the animals that inherited them. But complex traits take a long time to evolve and, more importantly, acquiring new characteristics is not like adding new apps to your phone — they are not simply bolted on wholesale to an organism. Hence, it wasn't the case that pelycosaurs acquired trait X, then therapsids got trait Y and mammals trait Z ... Parts of animals do not exist or evolve in isolation. So, while I will consider mammalian traits one by one, this will be done, always, with one eye on the fact that various characteristics often emerged and morphed in parallel, rather than one after the other. What we justifiably define as individual traits are nevertheless always bound together, and it is within these bonds that a meaningful notion of mammalian-ness can be found: though it's a bit hackneyed to say so, the whole is greater than the sum of its parts.

A final word on the duckbill

When Darwin inspected the platypus his host shot for him, he noted that its famous bill was quite unlike the hard, dry protrusions of England's museum specimens (indeed, platypus bills are made of very different materials from rigid bird beaks). And, credit where it's due, at the front end of the platypus, Everard Home did an excellent job of interpreting this bizarre structure. He wrote that the nerve running from the bill to the brain was 'uncommonly large', and suggested that the bill 'answers to the purpose of a hand, and is capable of nice discrimination in its feeling'. But, today, we understand that, while the bill does have 'nice discrimination', it also has a capability that no hand does.

We now know that when a platypus searches riverbeds for edible crustaceans and other morsels, the grooves in which its eyes and ears sit are pulled tight, and a valve of skin across its nostrils is likewise closed; the animal forages virtually blind, deaf and without a sense of smell. As it arcs its head from side to side, sweeping about the riverbed with its bill, that structure alone searches.

In the early 1980s, two German scientists found some unusual types of nerve endings in the platypus bill sparking suggestions that these animals didn't just have a regular sense of touch, but that they might also scan the water for electrical fields. Soon, a seperate team of German and Australian scientists placed batteries in tanks of water and saw that platypuses responded to them. Actually, they attacked them, plus they avoided electrically charged obstacles while bumping into inert ones. The scientists found that alongside 60,000 touch receptors, a platypus bill contains 40,000 electroreceptors. And that the large region of brain dedicated to input from the bill is arranged in exquisite stripes that alternate between receiving mechanical touch and electrical information. Echidnas seem to have remnants of this sensory system, but the platypus's duck bill is a wondrous sensory system seen in no other mammal: a reminder that evolution always balances conservation and innovation.

I've become stupidly fond of platypuses – enough to visit a stuffed one in a London museum more than once. Aside from their unique biology, I like that they swim about eastern Australia oblivious to all the confusion and consternation they've caused; unaware that people often call their anatomy and physiology transitional. They just keep on paddling and burrowing, diving for their food, experiencing their electrically sensed worlds, and once a year a female might lay a couple of eggs. That way, they carry on doing what they do, just as they have for millions of years.

CHAPTER THREE

Y, I'm Male

A male platypus signals his carnal intentions to a female by grasping her tail in his bill. Locked together this way, the pair engages in all sorts of watery acrobatics. Sometimes, she swims off and he, staying clamped to her rear, follows. If she turns 360-degree pirouettes, he twirls along behind her – the flirting couple cutting through the water like a furry corkscrew. There are reports that the coital deed itself is done with the male still biting his partner's tail. Whether this is true or not, a recently 'active' female can be identified by a bald patch on her tail.

When this routine results in the female becoming pregnant, she burrows 20 metres into a riverbank, builds a new nesting chamber, lines it with leaves, then lays her eggs – usually two. Having kept her eggs inside her for about 17 days (there's still uncertainty about exactly how long), she then incubates them for 10 days before providing the hatched infants – her 'puggles' – with milk for three to four months.

By contrast, once the aquatic erotica is over, the male paddles off and has nothing more to do with either the female or his offspring.

In humans, things aren't normally so stark, but still, by virtue of the Y chromosome that resides in each of my bodily cells, when it came to making babies, I had to do so with someone who had a twin pair of X chromosomes. And my failure to transmit that Y chromosome to my offspring means I am now the man of our house, a sole male with his partner and daughters. I am father, I am daddy, papa, patriarch, *pater familias*. Hear me ROAR!

Or something like that. Let me make a confession. When I sat in that hospital ward enraptured by the sight of Isabella suckling for the first time, the joy and the profound relief that engulfed me were spiked with something else, another layer of feeling, something unexpected and strange. I saw in that moment that there were certain things in this life that I would never be able to do.

Quite why such a thought had not struck me earlier in the reproductive process – say, at any point in the seven months Cristina had been harbouring a person in her abdomen – I will never know. All I can offer is that when Cristina had been pregnant, everything had been too abstract, a daughter had still only been an idea. In the hospital, feeding, Isabella was a person, a person whose fortitude and dignity humbled me daily. Watching her suckle, I wished desperately that I too could help her as Cristina was doing.

On the other hand, I probably shouldn't overstate this because subsequently, when my two daughters' hungry cries pierced the precious tranquillity of their parents' sleep, I typically clutched my side of the duvet and murmured that the girls needed their mum. Only when Cristina was away would I again lament not having a couple of inbuilt feeding and pacifying devices.

I immediately get nervous when I start talking about anything relating to how men and women differ. But any way you cut it, if you want to talk about mammals and what makes them unique you have to talk about the way they reproduce. And if you want to talk about how mammals reproduce you have to talk about males and females. When I talk about sex determination here, I am referring to the most elemental aspects of sex, whether the foetus develops testicles or ovaries and the most direct consequences of that. Full sexual differentiation is beyond this discussion, and gender identity is a nuanced and complex phenomenon lying even further beyond.

While this entire book owes its existence to the impression an errant football made on a uniquely male vulnerability, now as we enter the crux of having a mammalian baby, it becomes a story primarily about female bodies. No one trying to figure out the nature of platypuses cared much about the males; the animal's reproductive intrigue lay with the females. Mammalian reproduction is about sperm and egg meeting inside the *females*. It's about the *female* harbouring the resultant embryos as they grow, and about those embryos – except in monotremes – engaging in complex and dynamic relationships with their *mothers*. And it's about those relationships continuing after the young mammals are born, about those immature animals not having to feed themselves but having all of their dietary needs met by their *mothers'* mammary glands.

Therefore, central to the story of mammalian reproduction is how a straightforward duct that for aeons had simply passed eggs from the ovaries to the open sea was transformed by female mammals and their antecedents into a continuum of fallopian tube, uterus, cervix and vagina, and how females developed the ability to make milk. These innovations profoundly influenced mammalian life.

And what did males do while this was happening? They took *their* simple ducts for getting sperm from the gonads to the outside world, and stuck devices on the end for improving their aim.

Not that this is surprising: in over 90 per cent of mammals, a father's work is finished – like the platypus – the moment he's done ejaculating. The only contribution those 90-plus per cent of dads make to their children's welfare is the deposition of some chromosomes into an egg. And as for the males that do contribute to their offspring's development, evolution has generally sculpted their behaviour, not their bodies. Male bodies are mainly honed by *pre*-fertilisation challenges rather than *post*-fertilisation ones, that is, the task of getting *their* sperm to be the ones that switch events from pre- to post-fertilisation.

Later chapters will deal with how these various reproductive innovations came to exist. Here though, the focus is on how a mammal turns out to be male or female.

X and Y

My maleness is the result of an egg having been fertilised by a sperm that brought with it a Y chromosome. I am XY. My two offspring were created by sperm of mine that carried X chromosomes. My daughters are each XX. That's the coin toss of human sex determination.

To revise the basics: humans – nearly all of us, things occasionally go differently – have 46 chromosomes arranged in 23 pairs. Each pair contains a chromosome from mum and a chromosome from dad. For pairs 1 to 22, the twinned chromosomes are the same, bar some standard genetic variability between them. The final two are either a pair of X chromosomes or an X and Y. If it's two Xs, the person is usually female; if it's an X and a Y, the person is typically male. Eggs and sperm each carry only 23 chromosomes – one of each type – and because of the XX = female, XY = male configuration, every egg a female pops into her oviducts contains an X, while sperm are a 50:50 mix of those transporting Xs and those toting Ys. The sex of a human embryo is, therefore, decided by whether the sperm that successfully fertilises the egg carries an X or a Y chromosome. This system of producing a next generation that's half male and half female is as simple as it is effective.

In fact, it's so simple and effective, and X and Y chromosomes so fabled, that ever since I knew about it, I assumed it was the standard means by which you make a male or a female *anything*. This casual assumption was, however, an error, and quite a considerable one. The X and Y chromosomes that make humans male or female are entirely mammalian inventions.

Sexual reproduction is a very old process, and while its exact billion-year-old origins are a bit murky, its fundamental advantage over asexual reproduction is that sex shuffles a species' complement of genetic variation. It creates new organisms that explore the possibilities of combining in new ways the different versions of genes possessed by their mothers and fathers. It facilitates adaptation.

Reproducing this way doesn't, though, actually require two sexes. In fact, when you exclude insects (of which there are over a million species), about a third of all animals (vertebrates and invertebrates combined) are hermaphrodites. There are clear advantages to each member of a species having organs for making both 'male' and 'female' gametes (sex cells). First, such animals have a backup plan. If the animal fails to find a mate to mix genes with, it doesn't have to perish; it can fertilise itself. However, before having to resort to this, it would have had twice the chance of finding a mate in the first place, because in hermaphroditic animals, all members of the species are potential reproductive partners. Having separate males and females means only half your species can actually reproduce with you, which is quite a price to pay.

Then why bother? It seems that it goes back to the nature of the gametes themselves. Sperm cells and eggs are very different. A sperm cell is just a headful of DNA with an outboard motor and a tail, whereas an egg contains DNA plus all the energy and cellular machinery an embryo uses for its initial development. The two cell types, therefore, represent opposite ends of a spectrum of reproductive strategies. Sperm, being small, cheap and easy to disperse, give their manufacturers good odds of siring offspring. Conversely, eggs, being large, expensive and packed with resources, produce embryos that are more likely to prosper. It seems that early on in the history of sex cells, evolution deemed that being intermediate in terms of size and strategy was inferior, and the middle ground was wiped out.

The hermaphrodite, in being part-father, part-mother, must be a generalist. But animals can start to become specialised in producing either eggs or sperm – their bodies and behaviour developing traits that make them better at distributing sperm or at being careful with high-calibre eggs. When this happens, you end up with separate males and females, a scenario that has been – with the exception of a few hermaphroditic fishes and lizards – a fact of vertebrate life since its inception.

Besides halving the reproductive pool of a species, having two sexes also presents a fascinating problem: from a single genome a species must make two types of animals. There has to be a developmental switch that directs the growing body along one of two paths; a switch that, in the mammalian case, somehow involves X and Y chromosomes.

In addition to my misguided assumption that X and Y chromosomes were the standard way by which males and females were made across the entire biological realm, I was also off-target in my belief that the X and Y chromosomes were thus called because the X chromosome resembles an X and the stubby little Y chromosome looks vaguely like a stubby little Y. These terms actually stem from 1891, when Hermann Henking employed newly developed staining techniques to inspect the sperm of firebugs.

Before Henking's work, chromosomes had been almost invisible and the physical basis of inheritance a mystery. The stains he used were a revelation – chromosome literally means 'coloured body'. And if firebugs seem a curious choice, bulky insect chromosomes are actually much easier to visualise than the small, scrunched-up and considerably more numerous ones of mammals. The latter were indecipherable under Victorian microscopes and remained a challenge for decades.

Crucially, the firebug sperm Henking examined were not all alike. Half of them contained a curious little particle that looked like a small chromosome but didn't behave like the others. For its enigmatic nature, Henking named it the 'X element'.

The term resonated, such that in 1905 when Nettie Stevens, at Bryn Mawr College, Pennsylvania, resolved the chromosomal basis of sex determination by finding that male mealworm beetles had a pair of chromosomes that were different from one another, whereas females had a matching pair, these chromosomes became known as the X and the Y.

Stevens's seminal observation was important for two reasons: first because of its direct explanatory power in the field of sex determination; second for its wider implications for genetics. The twentieth century had begun with the rediscovery of the laws Gregor Mendel had uncovered by breeding pea plants – laws that suggested genetic particles underlying the inheritance of wholesale characteristics were passed on undiluted across generations – and biologists were fervidly applying these rules to humans and other organisms. However, genes were still purely hypothetical entities in need of a physical basis. While chromosomes' location within cells and their behaviour during sex and conception made them prime candidates for being the structures that contained or carried genes, discovering that an animal's sex correlated exactly with the chromosomes it possessed was exceptionally strong evidence in support of this idea.

Additionally, the XY system seen in beetles, and then in fruit flies, explained perfectly the confusing nature of inheritance of certain human diseases that disproportionately affected males.* Together these lines of evidence led everyone to believe that human sex was also determined by the same XY chromosome system.

And regarding this idea, things kicked along nicely in the 1920s. At the famous fly room at Columbia University, New York – the hub of the burgeoning discipline of genetics – the mode of action of the XY system was uncovered in fruit flies. Fly sex varied according to the number of X chromosomes an insect had: one X, it was male, two, it was female. This was ascertained by seeing that a mutant fly with an X but no Y chromosome – a so-called XO fly – was male. He was infertile, but the Y chromosome played no role in determining his sex. One X = male, an extra X flipped things to female.

In parallel, Theophilus Painter, at the University of Texas, turned his attention from insect chromosomes to mammalian ones. Painter cut his teeth on opossums, confirming that these marsupials had X and Y chromosomes; but he wanted human data to complete his study. Mammalian chromosomes were still stubbornly difficult to see, and choosing the right cell to look at was critical, as was the high-quality preparation of those cells. Consequently, the existence of the human Y chromosome was confirmed in testicular tissue rapidly prepared after its removal from inmates of a Texas state

---

* Some conditions, such as haemophilia, are caused by mutations on the X chromosome. Males who inherit a disease-causing gene on their sole X chromosome will have the condition. Whereas females are unaffected if they have a normal gene on their second X chromosome.

institution castrated for their 'excessive self-abuse'. In the paper announcing that mammals have X and Y chromosomes, Painter boasts of how quickly he got these men's testicles into his special fixative solution.[*]

Everything was cohering rather nicely. Who can blame people for assuming that X and Y chromosomes in humans, opossums and all other mammals would determine sex exactly as they did in fruit flies?

Hormones and chromosomes

Seeing that females had two X chromosomes and males an X and a Y was compelling evidence that these microscopic structures determined sex, but it also posed the question of *how* a chromosome might shape fate.

While Nettie Stevens and others made their genetic discoveries, other biologists were studying the sexual anatomy of female cows born after sharing a womb with a twin brother. So-called 'freemartins' had masculinised genitalia, with the degree of masculinisation seeming to depend on how much blood supply they'd shared with their brothers *in utero*. Speculation mounted that males released a chemical signal that altered their sisters' development.

This idea inspired Alfred Jost, in late-1940s Paris, to conduct one of the most significant experiments in the history of sex-determination research. Jost postulated that hormones from a male twin could not only stimulate masculine features in a sister but that the maleness of the male himself relied on these hormones. If Jost were right, masculine features wouldn't be an intrinsic and inevitable endpoint of being genetically male; the development of male traits would depend on the actions of male hormones.

---

[*] The X and Y, Painter got right, but his other legacy was to substantiate an earlier claim that humans have 48 chromosomes, a number that entered textbooks and apparently made scientist after scientist imagine an extra pair of chromosomes in the blurry images of human chromosomes that were available.

The likely source of these hormones suggested an obvious experiment. Jost would castrate some embryonic males and see what happened to them. Deciding that this could be done most easily in rabbits, Jost faced two main challenges. First, he had to acquire some rabbits. Usually, this would be straightforward, but in hungry post-war France, most people would rather have seen a rabbit in a stockpot than in a laboratory. Second, having got some rabbits, Jost had to master very fiddly surgery. Removing the gonads of rabbit embryos immature enough that they were still sexually neutral and replacing these ball-less bucks back into their mother's uterus with no collateral damage was no mean feat.

Mastering the surgery was, however, worth it. When Jost witnessed the birth of a healthy litter, he found that all the rabbits appeared entirely female. In the absence of testicular secretions in bodies containing X and Y chromosomes, nature had made female rabbits.*

As Jost started the 1950s figuring out exactly what testicles secreted to masculinise a male, genetics also made major leaps forward. Most obviously, Francis Crick and James Watson published the structure of DNA in 1953 and set in motion the quest to work out how this chemically banal molecule was the basis of inheritance. But geneticists also finally learnt how to reliably examine mammalian chromosomes.†

At last, medical geneticists could start to emulate the fruit-fly studies on which genetics had been built. In 1959, not one but three human conditions were concretely linked to unusual chromosomal inheritances. The first was Down's

---

* Despite Jost's experiment being done in rabbits, this type of hormonal induction of male sexual characteristics is not a mechanism unique to mammals; it is seen in many types of vertebrates.

† In 1956, the number of chromosomes in a human cell was finally corrected to 46. Only then, with the 48 Painter had seen in the Texan inmates unequivocally rejected, did numerous researchers confess to having already counted 46 but were too embarrassed to say so.

syndrome, seen to be the result of having a third chromosome 21. The second and third concerned patients with conditions that affected their sexual identities,[*] and they provided quite a surprise: humans were not like flies.

It was found that a person who was XO was not male – as such a fly was – but female; and that a person who inherited two X chromosomes and a Y was – despite having two X chromosomes – male. The conclusion was stark: if a human had a Y chromosome he was biologically male.

Combining Jost's findings with these startling genetic results seemed to say that a male mammal resulted from having both a Y chromosome and functional testicles. Sex determination in mammals, it seemed, might be as straightforward as a Y chromosome initiating male development by inducing the construction of a pair of testes.

In whichever mammals geneticists applied their newly acquired ability to see chromosomes, they found the same pattern: females XX, males XY. Except in one.

In the early 1970s, when researchers peered inside platypus cells, they found that even at this level these animals were quirky. There was no simple pair of sex chromosomes for them – oh no – instead they appeared to have five X chromosomes and five Y chromosomes. This sort of thing is precisely why I love platypuses. A female platypus is $X_1 X_1 X_2 X_2 X_3 X_3 X_4 X_4 X_5 X_5$ and a tail-biting male is $X_1 Y_1 X_2 Y_2 X_3 Y_3 X_4 Y_4 X_5 Y_5$.

How exactly this system worked was only revealed in 2004 when geneticist Jenny Graves and her team at the Australian National University showed that the five Ys and the five Xs line up in separate chains during sperm and egg production so that each chain collectively behaves like some sort of uber-chromosome. Graves is an expert in the genetics of marsupials

---

[*] These two people had, respectively, Turner's syndrome and Klinefelter's syndrome, conditions associated with a range of health concerns that vary in severity. Most pertinently, each syndrome results in infertility and issues with sexual anatomy, but in both instances the sex of the individual is unambiguous.

and monotremes, and both she and those animals would play pivotal roles in deciphering the genetics of mammalian sex determination. But for about 30 years, the $X_1 X_1 X_2 X_2 X_3 X_3 X_4 X_4 X_5 X_5$ and $X_1 Y_1 X_2 Y_2 X_3 Y_3 X_4 Y_4 X_5 Y_5$ genetics of platypuses sat on the shelf, considered just another monotreme eccentricity; a duck-billed take on the standard mammalian XY system.

From chromosome to gene

By the 1980s, genetics was ready to have a serious stab at establishing what exactly enabled the Y chromosome to trigger maleness. Attention shifted from people who were XXY or XO and on to a rarer group of individuals whose chromosomes presented a much greater conundrum: those 1-in-20,000 births that produced females who were XY and males who were XX.

To explain this, it was hypothesised that the Y chromosome contained just one gene that determined sex by instructing the embryo to convert its gonads into testes. In XY females, this gene would be lost or dysfunctional. And in XX males, it was thought that this crucial bit of genetic instruction had jumped ship from the Y chromosome and stowed away on an X.

David Page and his group at MIT in Boston spent a long time chasing this key bit of Y chromosome, and after a number of high-profile papers describing how they were closing in on it, they announced in 1987 that they had discovered a single human, male-determining gene. This gene was typically located on the human Y chromosome, but in XX men it was found on the X chromosome they'd inherited from their fathers. In XY women, the gene was missing from the Y chromosome. Page called it ZFY for zinc finger Y-chromosomal protein, and the paper they published in *Cell* was heralded as a triumph of brilliant genetics.

Now is probably a good time to return to my naive assumption that X and Y chromosomes, like my own, underpin maleness and femaleness in all sexually dichotomous species. This wasn't just wrong, it was really wrong – bodies can be made to be male or female in many ways. While the most common route is by the alternative inheritance of a genetic switch, it doesn't have to be. Not uncommonly, the factor determining sex is an environmental trigger: for example, whether a turtle or crocodile is male or female depends on the temperature it incubates at in its egg.

And when the switch is genetic, the genes used are by no means shared. To survey the full extent of sex-determining genetic mechanisms we'd have to look at plants, insects and all the other invertebrates that come in male and female iterations, but sticking to only vertebrates provides plenty of variety. In numerous groups, the males have different chromosomes and females a matching pair, like mammals, but phylogenetically these animals don't cluster together as a single group derived from a common ancestor. They're dotted about the vertebrate tree and we now know they use different genes and different strategies to sexually differentiate – sometimes there's a male-making master-switch (as in mammals), sometimes there's a dose-dependent system like the one in fruit flies.

Elsewhere, however, it's the females who have mismatched sex chromosomes and males that have a matching pair. Such an arrangement is seen in birds, snakes and certain others, where the female is said to be ZW and the male ZZ. The Z and W nomenclature is used merely to indicate that things are this way around; it says nothing about the chromosomes themselves.

Far from sex determination mechanisms being hard-wired and robustly conserved, as you might expect for such an elemental aspect of animal biology, they are remarkably prone to change. If you need further proof of this, visit Japan, where geographically separated populations of the local wrinkled frogs variously employ either an XY or a ZW system.

After Page had published his ZFY paper based mainly on human genetics and related placental mammals, he decided to beef up the case that this gene was the key switch in all mammalian sex determination. As part of this endeavour, he wanted to confirm that ZFY was on the Y chromosomes of marsupials. This seemed a pretty safe bet. Bar their Y chromosomes being a bit smaller than those of placental mammals, marsupials appeared to undergo sexual differentiation just as we did. Page sent Jenny Graves in Melbourne a short stretch of DNA that would stick to ZFY wherever it lay in the marsupial genome.

At the same time, Peter Goodfellow – a London-based biologist also working on sex chromosomes – likewise thought this was an interesting question and independently sent Graves a second probe for ZFY.

Graves gave the two DNA samples to her PhD student Andrew Sinclair and waited for him to bring her an image of a nicely stained kangaroo Y chromosome. Only Sinclair returned to say ZFY was on chromosome 5 – a bog-standard non-sex chromosome. Graves did what many a PhD supervisor has done when presented with a strange result: she told Sinclair to go back and check again.

But this instruction only resulted in Sinclair generating more convincing evidence that ZFY was not on the Y chromosome. There was no way a sex-determining gene could reside on a non-sex chromosome. Graves, Sinclair and Goodfellow wrote up a paper – with Page as a co-author – and sent it to *Nature*. Soon that journal's cover told the world that ZFY had been a mistake.

PhD in hand, Sinclair left Australia for London and Goodfellow's lab – to pursue the true sex-determining gene. And Page started afresh. Given that the foundation of exploring hitch-hiking genes in XX men had been sound, the search did not take long. In 1990, Sinclair and Goodfellow, working with Robin Lovell-Badge, published two papers – again in *Nature* – describing a gene named SRY for Sex Region of the Y. This time there was no mistake: SRY is a single uniquely mammalian gene, found on the Y chromosomes of both placental and marsupial mammals, that makes a mammal male.

Sometimes a single discovery unleashes a whole slew of fabulous research, and this was one such discovery. It's now understood that when the early undifferentiated, embryonic gonads of a mammal contain a complete Y chromosome, SRY is switched on and its presence ensures that those gonads become male. SRY does this by activating the expression of a network of other genes that activate further genes that guide testicle construction. The identities of about 30 of these genes are now figured out, but more unknowns remain. Many similarities exist between these genes and those used in sex determination by non-mammalian vertebrates. Switches may change over evolutionary time, it seems, but the wider testicle-inducing network is well conserved.

The DNA sequence of SRY is surprisingly variable across mammals, but it does the same thing in each species. Geneticists, for example, artificially inserted the goat version of SRY into XX mice and watched them develop (mouse-sized) testicles; testicles that, through the hormones Alfred Jost had studied, directed the XX mouse to become male.

To help complete the story, Graves's lab set out to do the obvious thing and work out which of the platypus's five Y chromosomes contained the monotreme version of SRY. Graves's previous work had provided evidence that bits of the platypus Xs looked much like the human X. No one anticipated any problems.

Only SRY wasn't there. After a decade of looking, Graves conceded that SRY didn't exist in platypuses and that it was absent from echidnas too.[*]

This conclusion was cemented when Frank Grützner and Frédéric Veyrunes, working in Graves's lab, looked at the platypus sex chromosomes employing technology that was being used to sequence the animal's whole genome. By that point, it was no real surprise that SRY was missing, but the results held another shock. The earlier indications that platypus X chromosomes looked like human Xs had been wrong. Genome sequencing revealed that platypus chromosome 6 –

---

[*] Echidnas have five X chromosomes but only four Ys, Y_5 having fused to another Y chromosome.

a run-of-the-mill non-sex chromosome – most closely resembled the mammalian X. If the platypus sex chromosomes looked like anything, they looked like the Z and W chromosomes of birds!

Graves describes this as a bombshell. First, it said they had no idea what made a platypus male or female. In birds, as mentioned earlier, things are the opposite way around from mammals: males have two Z chromosomes while females have a Z and a W.* It was perplexing that platypus sex chromosomes should resemble these but work the other way around. Still today, no one knows how platypuses turn out to lay eggs or bite tails. The best current candidate is a gene that determines sex in fish.

Second, these findings meant SRY had hijacked mammalian sex determination in the short span of time between monotremes branching off the mammalian tree and the arrival of the last common ancestor of marsupials and placentals. Sex determination by SRY was a uniquely therian mammal trait. Plus, this finding gave SRY and the Y chromosome a birth date: somewhere between 166 and 148 million years ago.

DNA as historical text

When Crick and Watson resolved the structure of DNA, they showed that it was both arrestingly elegant and frankly uncomplicated. Its elegance lay in its two strands running in opposite directions through and around each other. Its simplicity was in the confirmation that it really was just a long chain of only four chemical bases – the genetics alphabet of A, C, G and T.

Besides the critical insight of how bases always paired in the same way between the two strands, there appeared to be no rule as to what order these bases could be strung together in. Yet somehow this ordering and the information it held was the

* Birds use an entirely different gene to determine sex, and they're like fruit flies in that a double dose of this gene produces a male.

baton passed from generation to generation. Quickly, the question became: how does a one-dimensional sequence of DNA give rise to three-dimensional life forms? And soon there was a consensus that the sequence of As, Cs, Gs and Ts must direct the order in which amino acids were arranged in proteins, sparking a high-octane quest to crack the underlying code.

Francis Crick was instrumental in solving this riddle,* but he also pondered what else would emerge from a new understanding of genetics. Seeing that differences between organisms must lie encapsulated in their DNA and, hence, protein sequences, Crick (as he frequently did) fully grasped the implications. In a far-reaching 1957 lecture he said, 'before long we shall have a subject that might be called "protein taxonomy", the study of the amino acid sequence of proteins of an organism and the comparison of them between species.' Crick predicted that these sequences might provide 'vast amounts of evolutionary information'.

He saw that because changes in DNA sequence are how evolution yields new life forms, comparing sequences across species should allow a probability-based inference of life's history. Instead of examining how teeth, say, changed in form, you could instead look at differences in gene sequences to see how different lineages have separated over time. Hence, within a few years of its structure being determined, DNA added to its already impressive CV the role of a historical record of evolution (as we saw with INSL3 and its receptor in Chapter 1).

Nevertheless, having a good idea is one thing in science, making it a reality is another challenge. However, also in 1957, Emile Zuckerkandl – a French biologist whose nomadic academic life had been marked by few, if any, significant achievements – met Linus Pauling in a Parisian hotel to canvass Pauling for a job. Pauling had won a Nobel Prize for work on chemical bonds and protein structure, work that had aided Crick and Watson's pursuit of

---

* It is the sequence of three consecutive DNA bases that specifies which of 20 amino acids is placed next in a protein.

DNA's structure, and was now dedicated to studying the molecular basis of life and disease.*

At the California Institute of Technology (CalTech) in 1959, Pauling put Zuckerkandl to work on determining the fine-scale differences between haemoglobins extracted from different primate species. Pauling knew haemoglobin well; he'd previously shown that the blood disease sickle cell anaemia was caused by defects in this protein's ability to carry oxygen. When Zuckerkandl assessed the variability in the sequences of haemoglobin's amino acids, a very simple idea emerged. Zuckerkandl and Pauling proposed that if they knew the extent to which haemoglobin differed between two species and the amount of time – based on the fossil record – since those two species had last shared a common ancestor, they could calculate the rate at which molecular evolution had proceeded. If there were 10 amino acid differences in the haemoglobin sequences of two species that had diverged 100 million years ago, this would mean an amino acid substitution equated to 10 million years having passed. They had devised a 'molecular clock'.

The beauty of this clock was that if you knew how fast its hands moved, you no longer needed fossil evidence to date the divergence of species. Three amino acid differences between two species? That's 30 million years of separation. Instead of comparing the complex forms that living beings take, one-dimensional DNA sequences could be laid out side by side, the differences between them straightforwardly catalogued, and inferences made about the relatedness of the various species from which the genetic material had come.

It was an idea that changed everything. For the century since Darwin, genealogical relationships between living (and dead) species had been carefully deduced by comparing anatomy and morphology. Now, apparently, it could be done by taking DNA samples.

---

* As well as vigorously campaigning for peace and nuclear disarmament.

Some morphologists welcomed the new approach, seeing it as an invaluable adjunct technique, but many were suspicious of it. To be fair, you can imagine how strange it must have been. To compare species according to their three-dimensional forms, one must know them extensively and exactly. But here were some molecular biologists – who might never have seen the animals in question – claiming that their rows of plastic tubes filled with clear droplets of DNA-containing liquid held far more information than all that intimate anatomical knowledge.

To their credit, Zuckerkandl and Pauling were careful to consider caveats and issues with their approach.[*] However, there was no quelling the zeal with which molecular evidence was applied to evolutionary studies.

The infamous low point for naive molecular biologists with a taste for iconoclasm was in the early to mid-1990s, when it was twice claimed that DNA analysis proved guinea pigs weren't rodents. It was a suggestion that infuriated (and amused) morphologists, and was eventually shown to be due to mistaken interpretations of small data sets. But then in the late 1990s, improved techniques and new investigations of reams and reams of As, Ts, Gs and Cs entirely rewrote mammalian history – an event we'll return to in Chapter 9.

For now, though, we'll focus on the other major insight Zuckerkandl and Pauling had: one that wasn't about the same gene in different species, but about different genes in the same species.

Mammals don't just have one type of haemoglobin; they have four. And by comparing different human haemoglobins,

---

[*] Zuckerkandl and Pauling once wrote, 'Our best excuse for making this present evaluation is that it affords us the opportunity to point out why it is probably wrong.' Molecular clocks certainly proved to be considerably more complicated than being able to say that an amino-acid difference equates to x millions of years of phylogenetic separation. Today, there's extensive, frequently heated debate about how best to calculate rates of DNA sequence change and hence when lineages diverged.

Zuckerkandl and Pauling saw that the disparate proteins had almost certainly evolved from a single ancestral haemoglobin. The gene, they argued, must have duplicated to produce copies of itself, with each independent version then being freed to take its own evolutionary path.

It is now understood that new genes are very rarely built from scratch. In addition to genes mutating to incrementally change the amino acid sequence of proteins, larger stretches of DNA duplicate or otherwise get rearranged – whole chunks are deleted or thrust together in new combinations – and the ensuing possibilities of these new configurations are put through the sieve of natural selection.*

Where then did SRY come from? In 1994, Graves and her student, Jamie Foster, described a gene on the X chromosome that resembled SRY. It was called SOX3, and they proposed that SRY had originated as a variant form of SOX3. The mutation that began this conversion must have converted SOX3 into a gene that directed a developing body down the male route. This would have happened when the chromosomes that were ancestors of the current X and Y chromosomes were just a regular old pair of identical autosomes. Only when that fateful mutation occurred did the chromosome harbouring the SOX3/SRY ancestor become *de facto* a sex-determining chromosome.

Strange things can happen to chromosomes when they determine sex. This is because a female (in this instance) doesn't ever possess the male-making chromosome, and yet life must go on without this chromosome in that half of the species. The history of the mammalian Y chromosome is consequently a tale of atrophy and loss. First, bits of the Y chromosome flipped around so they couldn't interact with the X, and then they were lost. Now, the human Y chromosome harbours only four genes that were on the original X/Y ancestral chromosome. Its primary function is to carry SRY, it seems, and to provide some genes involved in making sperm. One recent study suggested

---

* And, of course, the traces of these duplications or rearrangements provide another window into biological history.

that only two Y chromosome genes are required to produce a fertile male mouse.

In view of all this degradation, it's currently a moot point as to whether the human Y chromosome is on an inevitable road to oblivion. Graves believes that it is, whereas others cite evidence from primates to argue that the diminutive chromosome has now reached a steady plateau.

It is, however, known that two mammalian lineages have abandoned SRY. Both are rodents. Certain mole voles lack the Y-chromosome genes, whereas Ryukyu spiny rats have attached some of their Y-chromosome genes to their X chromosomes but not their copy of SRY. Currently, a male-specific gene on chromosome 1 is a prime candidate for having taken over sex determination.

What is fascinating here is that the phenotype – the existence of males and females – persists, despite the loss of what, in nearly all mammals, is the crucial gene for creating that phenotype. Likewise, bodies coming in male and female iterations pre-dated SRY. This gene did not create sexual dimorphism; it simply became the means by which therian mammals achieved it. (SRY formed in just the right animals at just the right time to be carried on the winds of the mammalian radiation, so as to make mammalian sex determination genetically unique and almost genetically uniform.) This speaks to a primacy of the phenotype, and of how traits have lives independent of the specific genes that underlie them. A phenotype can persist while being achieved, over time, by different genetic means. Every generation, it's the living organism that emerges from the logic of DNA's four repeating chemical bases that matters.

And for us sexually dimorphic mammals, this construction process begins with males and females finding a way to come together, so to speak, to combine chromosomes.

CHAPTER FOUR

The Mammalian Birds and Bees

African striped weasels take over an hour. Lions take 10 to 15 seconds – but, when it's the season, they go at it 20 to 40 times a day. On average, humans take four or five minutes. Dogs get clamped together doing it. As do short-tailed shrews, with the otherwise inactive male being dragged around for 25 minutes. It can kill a female ferret to go too long without it.

Sex. Coitus. Intromission. The reproductively necessary liberation of male gametes in the vicinity of female ova. The fine art of lovemaking.

Robinson's mouse opossums do it suspended from a tree by only the male's tail, while a male stump-tail macaque monkey is most likely to be attacked by other macaques just as he's finishing doing it. Porcupines engage in it only if the female approves of the urine the hopeful male sprays her with. There is a species of mouse-like marsupial that once a year breeds *en masse* in tree hollows – the orgies are so ferocious, and levels of competition between males so intense, that all the males' immune systems collapse, and not one lives to see his first birthday.

All this said, most mammals still get down to business in the 'quadruped position'. The quadruped position is the one in which the female is on all fours – or lying down in the case of llamas – with the male behind, often standing on only his hind limbs. In the human realm, the ancient Romans called this arrangement the 'Position of the Wild Animal', the *Kama Sutra* lists it as the 'Congress of the Cows', and you might know it with reference to man's best friend. Notable exceptions to this positional monoculture include ourselves, who do it any which way we can (or, at least, try in our youth positions other than the two or three we eventually settle on). Bonobos, who also display excellent configurational variety, use sex as a greeting, a means of defusing tense situations and, seemingly, as a fun way to pass the time. Orangutans, who often favour the missionary position. And dolphins and whales, who are required by the anatomical reconfigurations that accompanied resumption of an aquatic life to adopt various front-to-front mating positions. But what's especially interesting about mammalian sex is the anatomical uniqueness of the body parts with which we copulate.

I'm going to keep this chapter light on personal anecdote – I think that's best for all concerned[*] – but I will say this. For the first 18 years that I engaged in sexual activity, I had three rules:

1 Don't catch a disease.
2 Don't get anyone pregnant.
3 Without breaking rules 1 and 2, have fun.

It was quite something to decide to drop Rule 2. One might think that this modification would have loosened the constraints on Rule 3, but instead 'foreplay' in the reproductive phase of my life was sometimes little more than a gruff reminder as I watched TV: 'Don't get too comfortable, I'm ovulating!'

We humans are so used to sex as a recreational activity, we forget what a serious business it is. The only rule of sex most organisms have is:

1 Reproduce.

Intimacy and fun are bonus add-ons for only certain lineages. Male wind-pollinated plants need only possess devices that allow their pollen to catch a breeze so that it might be deposited on a lady plant downwind. Poor things. And the fish from which we mammals ultimately evolved got barely any more familiar with each other. They bred, as most living

---

[*] You'll simply have to guess what those two or three positions are.

fish do, by spawning: males and females released their sperm and eggs into the water they lived in, and fertilisation occurred via random collisions within a marine miasma of sex cells. Such spawning is the original form of animal sex. But then, as the long journey from fish to mammal proceeded, sex became increasingly up-close and personal. First, internal fertilisation came along. This was a landmark event, albeit it is a mode of sex that numerous other animal lineages have also hit upon, even a good number of fish. Eventually, however, mammals became the only animals to copulate by the insertion of a penis into a vagina.

The uniqueness of this form of sex is owed to the females, for, in strict anatomical terms, the vagina is a purely mammalian invention. I once met someone so taken with this fact, she wanted to write a song called 'Only Mammals Have Vaginas'. I explained that this was only true according to a quite precise technical definition of a vagina, and other female animals had functionally similar organs, but she remained convinced it would make a catchy song.

What makes vaginas special is that their sole purpose – biologically speaking – is reproduction. Before there were vaginas, there were cloacas. A cloaca is a single posterior orifice that serves all its possessor's excretory and sexual needs – the name is derived from the Latin for sewer. Cloacas are found at the rear ends of reptiles, birds and, this is the correct term for describing a platypus's behind. You'll recall that this single entrance and exit route is why platypuses and echidnas were named monotremes. Although marsupials, it turned out, also have something approaching a cloaca where their various bits of plumbing converge.

In considering the nature of female mammalian genitalia, we return to the degree to which an animal's derriere is functionally compartmentalised, and that animal's consequent standing as a monotreme, bitreme or tritreme. And we're also back to how the ancestral oviducts that delivered eggs to protomammals' cloacas evolved into the multifaceted structures elemental to mammalian reproduction.

We'll start though with male anatomy, a subject that will take us back into pre-mammalian ancestry. For while mammalian penises are singular structures – no other males, for instance, urinate through their reproductive appendages – a lot of animals have phalluses*. And regarding vertebrates, there's been a longstanding debate as to whether mammals invented their mammalian penises from scratch, or whether they inherited a basic phallus from much older animals and, simply, made it their own.

Whence came the penis

In the Harvard Museum there is a series of microscope slides, prepared in 1909, which contain very thin slices of tuatara embryos. Tuataras are reptiles found only in New Zealand. Up to 80cm (2.5ft) long, adults look satisfyingly dragon-like. Once thought to be closely related to lizards and snakes, tuataras are now known to be the sole survivors of a lineage that diverged from other reptiles 200 million years ago. They are, it strikes me, the scaly equivalents of platypuses. And a male tuatara has no penis – a fact that has helped fuel the discussion about the origins of the mammalian penis.

As spawning doesn't require them, fish don't have phalluses. Among those fish who partake in internal fertilisation, a modified fin is often used to help get the job done. True penises only evolved in vertebrates after they'd made it to land. They're found in mammals, reptiles and birds,[†] but it's never been clear whether these various terrestrial lineages independently evolved sexual protrusions, or whether the penis evolved once and later diversified across the groups.

---

* A male extension of some sort typically evolves whenever sperm need transferring from male to female. For reasons discussed in the previous chapter regarding eggs being much larger and resource-heavy than motile little sperm, internal fertilisation occurs almost universally inside females. Seahorses are the most notable exceptions to this, where the females deposit their eggs in male pouches.

[†] A small number of amphibians have them too: see the next footnote.

Support for each lineage having made its own phallus comes from how profoundly different the various appendages are. Turtles, crocodiles and mammals all have single midline phalluses, but these organs are radically dissimilar. Lizards and snakes – collectively known as squamates – have double penises, sort of V-shaped protrusions, of which they rather disappointingly use only one branch at a time. And then there are birds: 97 per cent of bird species don't have penises, but among the few that do, there are ducks that famously have corkscrew-shaped phalluses longer than their bodies. Hence, the reproductive appendages of male terrestrial vertebrates don't immediately appear as mere variants of a single invention.

Additionally, the fact that tuataras and the avian 97 per cent mate via brief appositions of his-and-hers rear ends, in a process known as a 'cloacal kiss', gives further support to penises having had multiple origins. The efficiency of cloacal kisses shows that penises are not as vital to internal fertilisation as one might think, making it plausible that early terrestrial vertebrates bred without protruding coital aids and that various land-dwelling lineages only developed phalluses later. In this scenario, tuataras would represent a preservation of the ancestral state, just as platypuses' egg laying maintains a primitive mammalian trait.

The alternative model – whereby the penis evolved at the base of the terrestrial family tree, before diversifying or disappearing – gained support from investigations of the structure's development. Embryos of birds that were ultimately penis-less showed that genital outgrowth *began* just as it does in animals destined to develop full-blown male organs, only the growth then stopped and the tissue regressed. Also, mammalian, reptilian and avian phalluses had comparable initial stages of growth and used similar genes to direct this process. These shared mechanisms were presumably conserved from the initial penile innovators, indicative of a single phallic origin. But to really nail this idea, it needed to be known if male tuataras also lacked penises because they aborted a genital construction program.

If tuataras underwent some initial penis growth, it would suggest they're descended from ancestors that once had this organ, whereas zero growth would be more consistent with them representing an ancestral state that made no penis at all. *Wouldn't it be great*, went a fateful conversation among penis researchers at Florida University, *if we had some tuatara embryos?*

This was a few years ago, when Martin Cohn, a professor of the genetics of genital development, was chatting with his postdoc, Thom Sanger, and graduate student, Marissa Gredler, and it was assumed that obtaining such embryos was pure fantasy as tuataras are strictly protected: adults since 1895 and eggs since 1898. No appeal based on the evolutionary enigma posed by the penis was likely to sidestep the ban on using this species for experiments.

But Sanger had a surprise for his colleagues. While he was previously working at Harvard, a curator at the museum had tapped an old chest with his foot and said, 'That might interest you, it's full of reptile embryos.' Sanger, an expert on lizards, rifled through the box and was amazed to find in the back right-hand corner slides labelled with tuatara's scientific name.

Shortly after adult tuataras became protected in 1895 – and surely precipitating the protection of their eggs three years later – an Englishman called Arthur Dendy travelled to New Zealand and collected, incubated and dissected 170(!) tuatara embryos to study the animal's development. He sent four to the Harvard Embryological Collection where they were sectioned, mounted on slides, and probably left untouched for over a century. Sanger had been desperate to use them for something, and now he had a reason to.

Being among the world's foremost experts on penis development, Cohn's group knew exactly when to look for embryonic genital swellings. One Harvard embryo was too young and two were too old. But one was precisely the right age. The team scanned each slice of that specimen and built a 3-D digital reconstruction of it. Sanger was unsure what they'd see. He feared there would be some ambiguous bumps on the embryo's undercarriage that wouldn't allow any firm conclusions. But no – what he saw was unambiguously a pair of genital swellings, uncannily like those of the developing lizards he knew so well. These curious reptiles *do* – like all other terrestrial vertebrates – go through the initial stages of making a penis.

If we combine the overlapping genetic profiles of genital buds in birds, squamates and mammals with what these old microscope slides revealed, we have the strongest proof yet that mammalian penises are not our inventions but derivatives of a phallus that originated over 310 million years ago.[*]

I, tetrapod

With apologies to readers who anticipated a chapter on mammalian sex being non-stop titillation, I'd like to go on a brief detour into pre-mammalian history to look at where the ancient animals who invented the vertebrate phallus came from. Because these ancestors who first peered down at something sexual extending from between their, or their partner's, legs were unlike any animal that had ever lived before. They were the products of what's unanimously regarded as the greatest transition in vertebrate history: the emergence of land animals from fish.

People sometimes say vertebrates invaded the land, but this transition had neither the drama nor the wilfulness we usually associate with invasions. It took about 60 million years – from 400 to 340 million years ago – for a lineage of fish to become animals that could spend some of their time outside of water. Why this happened is uncertain but no fish ever set out to live on land. Rather, certain fish gradually moved into ever-shallower waters, evolving, along the way, a

---

[*] Sanger tells me there's one possible caveat to this story. A peculiar snake-like type of amphibian called a caecilian also has a phallus. It is *possible*, therefore, that vertebrate penises are even more ancient – invented by ancestors shared by amniotes and amphibians, and that non-caecilian amphibians lost their penises, as tuataras and most birds have. Currently, not enough is known about amphibian genitalia to be sure.

suite of traits that would eventually allow their descendants to fully step away from the ancestral aquatic home.

For example, one of the most notable of these traits gave land vertebrates their name. Mammals, reptiles, birds and amphibian are all tetrapods – named for their typically having four legs. Tetrapod limbs are modified fins, but this transformation was not driven by the advantages of walking on land. Instead, fins – which are typically delicate protrusions that steer bodies propelled forward by waves of muscular contractions passing along their main axes – were converted into structures capable of powering an animal's movement by predatory fish who ambushed their prey by thrusting themselves off the bottom of the bodies of water in which they lived.

How exactly nature transformed fins into limbs has been the focus of much research. By comparing the genetic programs that orchestrate the making of these appendages, developmental biologists have illuminated how fin-making genes were likely tweaked to add new elements to the originally simple structures – to produce limbs' upper and lower parts, and the ankles, feet and toes on their ends. The work is concerned with elegant cascades of genetic switches fundamental to all development, showing how different parts of a nascent leg switch on genes that establish their future identities. For example, a certain region will be the future ankle, and nondescript tissue will express certain genetic switches long before further genes are activated to initiate the process of actually converting that tissue into a joint between the leg and foot. I mention this because a penis, like a leg, is also an appendage that must be made to extend from an animal's body. Plus, female reproductive tracts, while not immediately leg-like, also have upper, middle, and lower components.

Lungs were another major innovation that permitted vertebrates to live on land. But, like legs, these breathing devices didn't evolve as part of a concerted effort to leave the water. Instead, the first fish that didn't rely solely on their gills to gather oxygen were simply making use of the fact that directly above the water in which they lived there was air brimming with oxygen. These fish likely lived in warm, oxygen-depleted water, so had much to gain from gulping the sky. At first, gas exchange occurred in the fishes' mouths, then later more elaborate internal surfaces evolved, these forerunners to lungs developing in fish that also had gills. Indeed, lungs are homologous to the swim bladders of most of today's fishes, not to their gills.

Lungs and legs weren't, however, nearly enough to allow an animal to scarper across the shores. For starters, for legs to function on land, the animal had to not sag between them. Water has buoyancy, an inherent upward thrust that counteracts gravity's downward pull, meaning a fish in water is essentially weightless. Air lacks buoyancy, so early tetrapods had to evolve stronger spines, ribcages and new abdominal muscles to win their battles with gravity.

Truly, life on land – or, more pointedly life surrounded by air – is so hugely different from life in water, that almost every aspect of the ancestral fish's body needed at least some updating before being fit for terrestrial living. To list a few examples, no longer able to suck meals in with water, tetrapods needed new jaws and swallowing mechanisms. To fruitfully manoeuvre their heads, early tetrapods also evolved necks. Most tetrapod sense organs had to adapt to the stimuli that triggered them behaving differently in air. Their circulatory systems, too, had to respond to the challenge of gravity. And the terrestrial organism had to deal with the way air can change temperature far faster than water ever does – the temperature on a beach can plummet or soar while the temperature of the adjacent sea barely shifts.

The sparse but fascinating fossil record of early tetrapods shows a lineage of animals changing everywhere from their ever-flatter snouts through to their progressively less finned tails. It was a succession of creatures that moved into increasingly shallower waters, adapting always to their immediate surroundings – to lagoons or to mudflats, say – never to the future. But when they reached the water's edge, plants and invertebrates had already set up camp beyond and, so, land held food that would help continue to pull vertebrates from the water.

Why did I want to take this little detour? First, in the context of this chapter, after reaching shore, terrestrial vertebrates soon invented sex as we know it. Second, terrestrialisation echoes what we said in Chapter 2 about mammals becoming mammals: no single trait definitively marked the arrival of a new type of animal. Third, the first true land animals were close to being the last shared ancestor of mammals and reptiles, the animals that lie at the start of this tale of mammalian-ness. And, finally – most importantly – despite all that had occurred in the evolution of tetrapods, the first terrestrial vertebrates were far from being masters of land-living. Vertebrates originated in water about 525 million years ago, after life had spent roughly three billion years as an exclusively aquatic phenomenon – water was what biology knew. The tetrapods that followed early plants and invertebrates onto land were doing something radically new. But while these animals had the ability to live on land, they were by no means masters of the terrain. For land animals, breathing, hunting and feeding, moving, coping with fluctuations in temperatures, and reproduction would remain arenas in which there was huge scope for innovation - these were the challenges our ancestors faced and met in their own uniquely mammalian way.

I, amniote

Soon after tetrapods had evolved functional legs, they split into two major lineages. One led to today's amphibians and remained more intimately tied to water; the other produced reptiles, birds and mammals, collectively called the *amniotes*.

Cohn and Sanger suspect that early amniotes concocted the basic phallus inherited by all subsequent terrestrial vertebrates, but this was only one of the major reproductive innovations associated with these animals. If the term 'amniote' is unfamiliar, amniotic fluid, in which a baby resides in the uterus, is more commonplace. By evolving new types of eggs within which their embryos were surrounded by fluid-filled membranes, early amniotes were able to lay eggs on land. Already capable of doing most of their business on land, they were now freed from having to find water in which to breed.

These eggs were laid in damp forests about 312 million years ago, where the original amniotes lived, looking much like small lizards do today. It was a time when terrestrial ecosystems were fast growing in complexity. The richness of plants growing on land had drastically increased, and with this, arthropods and insects had likewise diversified, providing rich food sources for amniotes.

In addition to overhauling vertebrate reproduction, these animals gained further independence from water via a number of bodily updates, of which I'll mention two here: waterproof skin and a new mode of breathing.

Water evaporates in air. Meaning a terrestrial animal with leaky skin is prone to this precious commodity drifting away, and so to dehydrating. The first tetrapods had skin somewhat thicker than the fish they'd evolved from, but amniotes went further and made the body envelope more complex and entirely impermeable to water. Amniote skin is a multi-layered structure with a tough outer coat of dying cells and, below this, a layer of water-repellent fat that stops water escaping.

Skin is a rather unheralded organ, but it's elemental to mammalian biology. In the reptilian lineage, skin became tough, scaly, and, in one group, feathery. Whereas in mammals, it remained softer and retained its ancestral secretory glands. And it's from such glands that some of Mammalia's most distinctive features sprouted: hair most obviously, but also means for making sweat,* scents and a particular nutritious white liquid.

Amphibian skin – which also contains secretory glands – has never been waterproofed because these animals breathe

---

* Sweating is an active and tightly regulated release of water when the body calls for it, not a leak. It has made water evaporation a blessing rather than a curse.

through it. Oxygen dissolves in amphibians' moistened outsides and passes into their bodies. This is handy for breathing underwater, but it also supplements amphibians' primary means of gas exchange, which isn't as powerful as that of amniotes. Amphibians breathe – as early tetrapods would have – by using their cheeks and mouths to pull air in and out of their lungs. While its persistence for well over 300 million years says this works well enough, amniotes developed a stronger ventilation mechanism. Amniotes breathe by directly expanding their lung-harbouring chest cavities. This manoeuvre – initially performed solely by lifting the ribs – lowers the chest's internal pressure and draws air in. Powerful breathing was essential for mammals later evolving increasingly oxygen-hungry metabolisms, but they would still have to find ways to make the system even more effective. Evolving chest-based breathing also had an interesting collateral benefit for amniotes – it allowed their lungs to be further from their mouths, which appears to have freed up more space for larger nerves to run to their front legs. Hence, amniotes could develop much more sophisticated front limbs than those found in amphibians. Another example of no trait evolving in isolation.

OK, let's get back to reproduction. Early tetrapods bred – as their piscine ancestors had and as their amphibian descendants do – by spawning, relying on water for reproduction and retaining a larval life stage. Only once the tadpole-like larvae had consumed enough calories to power their metamorphosis did adult tetrapods emerge who were capable of departing the water.

The eggs amniotes evolved can be likened to private ponds. They work by the mother enclosing her egg cells within shells that also contain sufficient water, energy and building materials to cater for the development of that egg into an independent organism.

In addition to the shell, the evolution of the amniotic egg involved the development of three new membranous outgrowths – structures that were necessary for the embryo to survive inside its fluid-filled egg. One membrane grows to surround the embryo and yolk; another – the amnion, containing the amniotic fluid – encloses only the growing embryo; and the third extends from the embryo's gut to have the curiously paired functions of disposing of an embryo's nitrogenous waste and breathing for it. The extension of this third membrane from the gut is why we mammals end up with belly buttons, as one day these membranes would be converted into the mammalian placenta (more of which in Chapter 6).

The evolution of the amniote egg over 100 million years before birds evolved often gives biologists a good chuckle at the fact that the egg definitely came before the chicken. However, these biologists are at a bit of a loss as to why exactly the egg evolved. The usual idea is that amniotic eggs could be laid on dry ground, facilitating amniotes' assumption of more fully terrestrial lives. But while this is almost certainly part of the egg's advantage, some living amphibians and invertebrates can lay their non-amniotic eggs on land. Instead, it may have been that amniote eggs offered greater mechanical support to the embryo, or that they could be made larger so as to produce more substantial offspring – the beginnings of mothers laying fewer but bigger eggs; a move to quality over quantity.

One thing is sure, however: these eggs would have evolved in animals that were already undergoing internal fertilisation. Amniote eggs need to be seeded by sperm within the female reproductive tract before being enveloped in shell, which means that while the egg may have come before the chicken, the cock almost certainly came before the egg.

As natural selection cannot see into the future, early amniotes could not have evolved penises and internal fertilisation in order to lay the groundwork for a future renovation of the entire way in which their babies were made. These animals were breeding in or around water, so penetrative sex must have evolved for other reasons. People who study the acquisition of internal fertilisation by fishes tend to think the males favour it over spawning because it gives them higher fertilisation rates and, perhaps, more certainty of paternity than waiting for females to release their eggs before ejaculating on them. Unlike fish, however, who repurposed fins for sperm transfer, finless amniotes had to invent a whole new dedicated organ.

Working out how this was achieved returns us to legs. From the research that mushroomed from interest in the evolution of limbs, Martin Cohn – who sent Thom Sanger to examine the midriffs of embryonic tuataras – applied the lessons of limb developmental genetics to penises. As noted, penises are, like limbs, three-dimensional appendages that must coordinate the development of a left and right, front and back, near and far ends. Through the early 2000s, Cohn found that a number of the signalling molecules and genes that achieve this for limbs are also employed in phallus construction. Amniotes didn't repurpose a body part, but they did repurpose a genetic toolkit for building appendages.

The link between legs and genitals was extended in 2014 when Patrick Tschopp and colleagues in Cliff Tabin's lab at Harvard showed which embryonic tissues phalluses grew from. During early development, it's possible to label specific cells fluorescently so that all their cellular progeny remains fluorescent, meaning you can see where the cells arising from the originally labelled cells pop up in more mature animals. Tschopp and co-workers, including Sanger, injected the cells that would produce limbs in lizard embryos and found that later the nascent phalluses of these creatures glowed green. That is, the lizards' genitals were made from cells descended from those that made legs – it was an observation that made much sense of the fact that lizards and snakes have double penises.*

---

* Comparing the specific genes switched on in lizards' legs and their genital swellings revealed a remarkable overlap. What's more, snake penises genetically resembled lizard legs too. Although snakes have dispensed with legs, they retain embryonic limb tissues from which to craft their reproductive organs.

These early amniotes inventing the penis meant that one day a peculiar species of intelligent apes would have something to laugh and fret about and to hide from public view. But what's crucial about the phallus is that we now have marital beds rather than matrimonial bathtubs. The penis changed the way amniotes had sex, thereby initiating a revolution in how they reproduced. Fertilisation, then development, ceased to happen in the great wide world, and became dramas that played out inside the mother.

Back to mammals

When Tschopp and co. investigated the developmental origins of penises in mice, they found that the mammalian lineage had transitioned from making penises out of limb precursor cells to using the precursor cells of their tails. Surprisingly, it seems the cluster of cells that organises genital development moved during mammalian evolution and now instructs entirely different cells to make the penis. Why exactly this happened is unclear, but this developmental quirk added to a small but notable list of peculiarities of mammalian phalluses.

As mentioned earlier, male mammals are alone in this world in using their members to urinate. This was a fairly late-stage development, as male platypuses don't do it. Platypus urine appears penis-bound when it leaves the bladder, but at the last moment it is diverted down a passage at the base of the flaccid phallus. It's only when an excited male has a female's tail clasped in his bill that the mechanics of platypus tumescence forces the premature exit route shut – these hydraulics, protrude the penis from the cloaca where it usually resides and ensure that semen passes along the penis's full length. And despite Everard Home's confusion over his female specimen, in 1802 he correctly inferred these separate seminal and urinal exit routes in his pickled male.

While urinating through a tube has definite late-night advantages for a modern-day human male who's drunk too much, it is hard to find a genuine biological explanation for it. If penile urination was a serious biological advantage, you'd expect female mammals to do it too. In certain female mammals, the urethra does run through the clitoris, but only rarely is this female equivalent of the penis an enlarged structure.*

While we're on this subject, mammals are the only amniotes who produce voluminous amounts of urine in the first place. Reptiles evolved a mechanism for excreting their nitrogenous waste as powdery white uric acid – think bird poo – a slightly costlier means of getting rid of ammonia, but better for conserving water than the urea-based system that mammals retain from fishes.

The other thing about mammals peeing through their penises is that mammals are the only amniotes with an enclosed tube running through their phalluses – all others ejaculate along a groove (sometimes, admittedly, a fairly closed one) running along the side of their organ. Such urethral grooves are evident in developing mammalian penises but are enclosed before birth to form the familiar tube.

Finally, the way mammals get erections is also pretty unique, although turtles convergently hit upon using blood accumulation to achieve rigidity. Other vertebrates use lymph fluid either alone or in combination with blood. The Ancient Greeks and everyone else until the Middle Ages thought human erections were achieved by an accumulation of air. The man who put this frankly ridiculous idea to rest has so much else on his CV that this achievement rarely gets a mention.

In 1477, in Florence, Leonardo da Vinci watched the dissection of a recently hanged criminal. According to

---

* Most famously, female spotted hyenas have 18cm (7in) clitorises, or pseudopenises as they're known. This hypertrophied organ is typically accredited to the massive levels of testosterone that circulate in the blood of these highly aggressive animals. Remarkably, the hyena's vagina passes through this structure, meaning that sex involves the male inserting its regular penis in the female's pseudopenis, and more dramatically the female gives birth through it, which is perilous for the young, especially the firstborn.

Leonardo, men executed this way often died appearing aroused. Having seen first-hand that the criminal's member was tightly filled with blood, Leonardo wrote his theory of the haemodynamic erection. He noted that even if the entire body were full of air, there wouldn't be enough to make the penis as 'dense as wood'. He supplemented his observations by reminding his readers that 'besides one sees that an erect penis has a red glans, which is a sign of inflow of blood'.

All this talk of 'the mammalian penis' might make it sound as if all mammalian willies are alike. This, however, is not the case. As a keen goalkeeper for a good chunk of my life, I thought my time served in male changing rooms had – amid the coyness, exhibitionism, and singular unspoken etiquette of those places – taught me about penile diversity. But oh, no.

According to the usual ranking criterion, mammalian phalluses range from shrews at about 5mm (0.2in) to the blue whale whose member is so considerable it has its own Wikipedia page. This page reports a length of around 2.4 to 3 metres (8 to 10 feet), but quite reasonably says that it is 'unlikely to be able to be measured during sexual intercourse'. Size alone, though, doesn't come close to capturing what evolution has done below the waist of male mammals. I recently surveyed mammalian penises much as I watch horror movies – constantly *wanting* to look away … Tomcats' penises, for example, are covered in spines, spines much larger than those on rodent and chimp penises. These protrusions are thought to induce the female to ovulate, which in cats happens only after sex. A ram's penis, by contrast, has what looks like an inside-out urethra protruding from it so that it sprays semen around inside the ewe. A pig's erect penis resembles its curly tail, turning a helix due to its two disparately sized erectile chambers. A walrus penis contains a 60cm (2ft) bone called a 'baculum' or 'os penis'. Indigenous Alaskans frequently collect them and carve them into decorative knife handles and the like. If these bones seem weird, know that most mammals have them – including primates – making it surprising that humans don't.* What the human penis does possess, however, is a glans that extends something like a mushroom cap. It's been proposed that this evolved to act as a plunger that, during pre-ejaculatory thrusting, sucks out any deposits left by previous encounters.

Marsupials and platypuses have twin-headed penises, which we'll return to presently. But echidna penises have not two, but four heads, which are called rosettes. This is strange enough, but when the echidna is aroused two heads retract and only two are functional.

Finally, I have to talk about tapirs. Primarily, because I'm haunted by a video I watched of a tapir. Its penis was long for the animal, but more than that, it looked like an independent organism. In fact, it reminded me of the creature that emerged from John Hurt's chest in the film *Alien*. And then there's Jack. Jack lived at San Francisco zoo, and Jack stood on his penis. His keeper reported that despite her best efforts, the penis 'turned purple, then black, then it atrophied. Then it fell off, and he ate it.'

What is one to make of all this (other than to be glad I never kept goal for a Mammal All-Stars football team)? One thought is that the mammalian penis is like many mammalian traits in being widely and wonderfully adaptable. But then, phalluses generally are said to be the most rapidly evolving structures in all of the animal kingdom. What's important is why this is so. The variety used to be attributed to a boyish arms race to have a penis better than your rivals, however, understanding has now moved beyond this notion. Sexual reproduction is a nuanced process and a process that must serve both participants. Science now appreciates that females are not as passive as people used to think.

---

* Whales and elephants also get by without them, meaning that the walrus bacula are among the largest. Curiously, the baculum has evolved and disappeared multiple times in mammalian history, and there are protracted and ongoing debates about what purpose it serves.

Vaginas

There's a problem with vaginas: people don't study them enough. In 2014, Malin Ah-King, then at Uppsala University, Sweden, and her colleagues systematically surveyed whether evolutionary studies of genitalia had focused on males, females or both. And the results were shocking. Of 364 studies of animal genitalia, 44 per cent considered males and females, but 49 per cent studied only males while just 8 per cent looked at only females.* The authors ruled out laziness, stating that 'the persisting male bias in this field cannot solely be explained by anatomical sex differences influencing accessibility'. Instead, they saw it as a reflection of an erroneous but enduring belief that males are the dominant players in sexual exchanges and that female genitalia are invariant and uninteresting.

But females do not want just any male to fertilise their eggs (some things sound so obvious when you write them out). And the history of sexual intercourse and the dynamics of genitalia evolution present an intricate interplay between males and females, of adaptations and counter-adaptations. Sure, penises evolve quickly, but not solely because of some locker-room competition; it's also because females develop mechanisms that serve themselves. Females evolve ways to ensure that conception occurs by their eggs meeting the best-quality sperm possible.

In papers with titles like 1993's 'Why do females make it so difficult for males to fertilise their eggs?' a new view of sex emerged in the 1990s. While males may compete among themselves to gain access to females, female bodies frequently contain means for giving them some control over who fathers their offspring, most of which operate post-coitus. Getting so far as having sex is not enough then to become a father – a successful post-ejaculatory fate for sperm is by no means assured. I won't go into the details of how females can

---

* While only 27 studies were concerned with mammals, the bias towards males in those studies was stronger still. I have no list of interesting mammalian vaginas. The research isn't there.

manipulate the success of deposited sperm – it is seen across all animals – but as examples, a female may store sperm to increase competition between them (a trick that can also allow females to control when conception occurs). Females can fail to prepare their uteri for implantation, as has been seen in rats and hamsters. The molecules that mediate the physical interaction between a sperm and an egg can battle to ensure a good match. The female immune system can be hostile to sperm. Female rodents and geladas – close relatives of baboons – have even been seen to terminate pregnancies when the arrival of new males endangered the future safety of their current foetuses.

In fact, in early amniotes, females making sperm swim further to reach their eggs might have been an essential step in the evolution of the amniotic egg. For the mother's body to prepare such eggs for laying after they've been fertilised, conception must occur high up in the reproductive tract.

Female carefulness is born of the differences between eggs and sperm (as discussed in the last chapter). A female invests significantly more resources in her gametes than a male does in his. And, as coming chapters will detail, when a female mammal commits to mothering offspring, the investment required is enormous. Barely any aspect of sex evolves unaffected by the fundamental mismatch between male and female investments.

Thankfully we seem to have turned a corner regarding research, and contemporary studies increasingly consider sex according to these dynamics. Now, it is appreciated that it's not enough to point and stare at male parts. What needs to be known is how the ewe's reproductive tract is adapted to the bizarre ejaculatory anatomy of the ram – or if an everted urethra is a response to female innovation; what the sow does about the corkscrewed tip of a boar's penis that helps lock him in place; and if female cats' ovulation after sex is provoked by a barbed phallus, might she *not* ovulate if a tomcat is not to her liking? I have no idea what a female tapir is to do with her partner's monstrous device, but, surely, she does something …

Now, though, let us travel further back to ask how the uniquely mammalian vagina evolved in the first place. To address this question, we are again served well by looking at how this structure develops. And the first thing this perspective reveals is that the vagina is not a single structure. A vagina is made by joining together upper and lower parts, which have separate developmental origins – the top part is a modified oviduct, whereas the bottom part is derived from the cloaca.

Yes … after all my smugness about being a therian mammal with external plumbing way more sophisticated than that of an animal with a cloaca, it turns out we all (males included), for a particular phase of our very early lives, have one. Embryonic mammals all, briefly, have rear-end cisterns in which their rectums, urethras and reproductive ducts terminate (albeit, while this is the case, the embryo is not sexually active, not eating anything, and the placenta is taking care of urinary matters). This arrangement makes more sense when you bear in mind that originally urine, faeces and gametes were all products that needed to be expelled – the female reproductive tracts of animals that breed by spawning are not receptive structures.

While the cloaca no longer persists into mammalian adulthood, it remains a pivotal developmental structure – it is both a cavity and a source of tissue and signals that orchestrate the construction of both male and female external genitalia.* The key event in the formation of the lower part of a vagina is the partitioning of the cloaca into front and rear compartments.

Imagine a woolly bobble hat. This is the cloaca. On the back of the hat, a substantial tube flows into it – this is the hind gut. And on the front half, the urethra and the oviducts enter the hat. (You may not want to wear this hat. In fact, for the next part of this thought experiment, you can't be wearing it.) What evolution did was to reach into the hat and grab the bobble and pull it downwards, to divide the internal space of

---

* It is a shift in the position of the cloaca that appears to have switched the origin of the penis from limb to tail tissue in mammals.

the hat in two. These rear and front parts are then sealed off, some sort of biological needle and thread making this dividing wall watertight. This way, the rear part becomes an extension of the hind gut, while the front part allows the passage of babies and urine out into the world and semen into the reproductive tract. It's called the urogenital sinus.

This gets us to a bitremic female. But we need tritremes, right? Well, actually, yes and no. Female humans are, in fact, kind of unusual in this regard. The majority of female mammals are not tritremes. Most female reproductive tracts go no further than this two-way division – separating the urinary and reproductive tubing from the solid-waste disposal unit is sufficient. A further parting of the urethra and vagina so that each exits the body individually occurs only in certain rodents and primates.

The upper part of the vagina – about the upper third in humans – is made of the terminal parts of the oviducts. The oviducts are the evolutionarily ancient tubes that gather eggs from the paired ovaries and transport them to the exterior. In humans, these two ducts fuse at the midline so that two oviducts – or fallopian tubes – terminate in a resultant single uterus, with a single cervix atop a single top part of the vagina. However, there's actually a good deal of variety in the extent to which the oviducts are joined across mammals. The human arrangement is among the most highly fused. This is a 'higher' primate thing; the wombs of most other mammals show quite clearly their derivation from twinned oviducts. In rodents and rabbits, there are separate uteri in the two oviducts, each with its own cervix. Then in deer, horses and cats, for example, two distinct wombs share a cervix, while elephants, dogs and whales have single uterus but with clear left and right horns at the top.* This sort of variation could make a person think that there's a series of intermediate fusional forms ascending toward the most perfected human condition. But that's a notion as dated as Queen Victoria's frocks.

---

* Some women have 'bicornuate' wombs like this.

Like rodents, marsupials also have two separate wombs, but they take things even further. Marsupials have three vaginas – two for letting sperm in, and one for letting their joeys out. The story of the first two evokes again the historical contingency of evolution, while the third is a total mystery.

In placental mammals, the ureters – the urine-carrying tubes that run from kidney to bladder – pass around the outside of the developing oviducts, which means that the oviducts are free to fuse in the middle. But in marsupials, the ureters descend between the oviducts, meaning that the reproductive passages cannot fuse. Hence marsupials have two vaginas that arch around either side and meet only in the cloaca. This twinned arrangement is almost certainly why male marsupials have two-headed penises.

Why the youngsters can't then exit the same way daddy's genetic contribution got in is unclear. But they don't. Instead, to give birth, a pregnant marsupial develops a third vagina straight down the middle. The discoverer of this medial vagina was the Enlightenment's pre-eminent explorer of surprising Australian reproductive tracts, Everard Home. Home did this while dissecting a kangaroo in 1795, but like the intricacies of the platypus, the full nature of female marsupials' nether regions took nearly a century to truly resolve. In the intervening 100 years, some anatomists agreed with Home, and some thought he'd imagined things. But in 1881, Joseph James Fletcher and Joseph Jackson Lister examined a series of kangaroos whose reproductive status they knew and showed that the medial vagina is formed specifically to give birth through. In some kangaroo species, it remained open after the firstborn had passed through it; in others, it resealed and reformed with each round of births, the latter scenario being the standard *modus operandi* for marsupials.

We'll return to marsupial reproduction in the next chapter. For now, a final note on the oviductal part of the vagina, and a nod to developmental biology. The oviduct in its most elemental form is a tube with a funnel on one end for catching eggs and an exit on the other for releasing those eggs. In amniotes, the evolution of this duct is a story of regional modification. Different areas of the oviduct first became specialised for providing an egg with albumin and then with a shell. In mammals, the shell gland became a uterus and a cervix, and the unique vagina evolved. In early development, long before each of those regions becomes distinct, a handful of genes, closely related to those that pattern legs and penises, are switched on, a different one marking each of the separate areas.

If Chapter 3 viewed DNA as both the core instruction manual for making organisms and a logbook of evolutionary change, the developmental processes by which organisms are made represent a way for biologists to examine how evolutionary changes in form and function have occurred.

CHAPTER FIVE

The Next Generation

In the long hot summer of 1976 – earlier than intended – a small motile cell from my father attached to an ovum tumbling down one of my mother's 21-year-old fallopian tubes. The sperm ended its long journey by burrowing through the egg's coating and depositing its 23 chromosomes inside. With that, its job was done. The fertilised egg split in two, four, eight, then sixteen … eventually, there were trillions of cells. When there were 16 cells, there was very little to distinguish one from another; when there were trillions, there was me.

According to a recent estimate, an adult human consists of approximately 37.2 trillion cells. A sperm and an egg conjoining is a landmark event, but making a person requires much more work. To get 37.2 trillion cells requires at least 37.2 trillion-minus-one cell divisions – that egg and its progeny divide an unimaginably large number of times.* And while the cells multiply, they must also undergo many transformations – they interact, assume different identities, travel, organise themselves into different tissues … Ultimately, cells must somehow form everything from penises and vaginas to heads, shoulders, knees and toes.

Building an animal is such a fantastically complicated process, it's almost funny that no sooner is it done than the mature creature's first goal is to start the whole process over again. Egg meets sperm – animal is made, animal develops to sexual maturity – egg meets sperm … Always, this procreational loop.

---

* In reality, many more divisions happen, because development proceeds by making more cells than are necessary and disposing of those that are not in the right place at the right time.

Perhaps, when we hopelessly try to grasp how many millions of years old life is, it's actually more dumbfounding to think how many ancestors we have – how many millions and millions of times this loop has looped.

For most of human history, no one had any idea how an animal developed. After Antonie van Leeuwenhoek invented the microscope and discovered sperm in the 1660s,* he became convinced that the head of each sperm contained a miniaturised animal that had only to grow. Others believed that the egg held all the mystery. 'Sperm in, baby out' is pretty close to alchemy, isn't it?

Today, developmental biology ranks among the most elegant and astonishing of all the sciences. From a blur of cells an animal emerges in a marvel of self-organisation, and we now have a good outline of how this happens. And this knowledge has relatively recently placed development back at the heart of evolutionary biology. Nothing about limbs and genitalia makes their construction uniquely useful for inferring their evolutionary history; with seemingly all body parts, appreciating their development can help increase evolutionary understanding. Knowing how bodies are made illuminates how different bodies *might* be built, and how current bodies might once have emerged from previous ones. It closes the gap between genetic change and final form.

However, being in a state of development – being immature – is also dangerous. The young are incomplete and incompetent. They are fragile, dehydrate easily, starve quickly, rapidly get cold or overheat. They have limited feeding options, they are easy prey. Immaturity is perilous. If you plot the likelihood of

---

* I love the fact that the man who invented the microscope discovered sperm – how long did it take Leeuwenhoek to think, 'I know what I'm going to look at on this'? No, I shouldn't be so crude: the nature of seminal fluid was profoundly mysterious at the time and Leeuwenhoek actually had to be cajoled into working on it, worrying about how he'd be perceived. He claimed, also, never to have defiled himself for this work, always using the natural by-products of conjugal interactions.

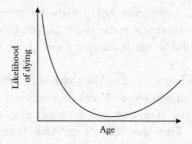

dying against age for any animal, you see that mortality is at its highest at very young and very old ages; the peril of immaturity and senescence flanking a plateau of relatively high survival.

Therefore no matter how glorious the end product of development is, this is not enough – whether we're talking elephants, tigers, rabbits, sticklebacks or daisies – evolution must sculpt the means for successfully passing through every developmental stage if the species is to keep on keeping on.

I, multicellular organism

Development is a problem only faced by multicellular organisms. Bacteria divide in two, and two new bacteria go their separate ways. I'm not sure if the advent of multicellular life gets the attention it deserves – compared to, say, interest in the ultimate origins of life or the emergence of humans. Yet multicellularity was a genuinely radical innovation. For the first two and a half billion years of life existing on Earth, there were only single-celled organisms: an organism was a cell, and a cell was an organism. Which is to say, for the first two-thirds of earthly biological history, cells found no value in clubbing together. The benefits of communal living, one must conclude, were difficult to stumble upon, or life was too different then. Either way, compare, if you will, a Siberian tiger to an *E. coli* bacterium. They are very different. No single cell has ever snapped the neck of a deer, and no large cat has ever lived in a human intestine. But these are superficial differences. Biology is so chained to its core mechanisms that multi- and unicellular organisms must still do the same basic things: the bacterium and the tiger both need oxygen and food to burn, both must balance their water content, and both have to excrete their metabolic by-products; each has to avoid or respond to life-threatening situations, and each must reproduce.

The big difference is that the single-cell bacterium has to meet these demands by itself, whereas the tiger is made up of lots of different cells that are very good at doing certain specific tasks. Tiger gut cells are great at absorbing nutrients from venison, but a retina lined with tiger gut cells would result in a blind tiger. When life went multicellular, that was the fundamental shift – cells within a commune started to specialise, and different cells started to contribute different things toward a common good – competition was supplanted by cooperation.

Imagine the earliest troupes of cells that stuck together in meaningful ways. They would have amounted to pretty amorphous clumps, the cells gaining some simple advantage from their togetherness. In such a setting, we must imagine, constituent cells started to carry out distinct tasks. Perhaps the cells on the outside became a little tougher than those in the centre. Then, maybe, inner cells – freed from the demands of being tough – did something that reciprocally benefited the outer cells; perhaps they performed a certain metabolic function more efficiently.

At first, these clumps were probably pretty anarchic. Cells would not have been programmed to be on the outside; they would simply have toughened up if they ended up there. Over time, however, increasingly exact modes of development evolved. The identities of cells – which were all clones, carrying the same DNA, and whose characteristics were determined by which subsets of their genes they switched on or off – would have become more rigid. Signals between cells would have started to ensure a more stereotyped arrangement of different types of cells; successive generations would have started to look more alike. Growth and development became a process that multicellular organisms went through.

Today, we take it for granted that animal bodies are made up of different parts, but every one of those parts had once to evolve out of a previous state in which it didn't exist. Hearts, brains, kidneys, eyes … all of these were once invented from scratch. And it's interesting to think briefly about why they evolved – some represented a simple divvying up of function, so that specialised cells could do specialised jobs. Sensory organs are wonderfully efficient at detecting things, much better than a single cell could ever be. Skin is an excellent outer barrier. An enclosed gut is great for holding food and plundering all the nutrients it contains. But other innovations appear to be adaptations to being multicellular in the first place. A single cell doesn't need a circulatory system – only when cells don't face the oxygen-containing outside world do they need oxygen brought to them. Then, if the blood also becomes a dumping ground for waste products, it needs organs that filter and detoxify it. Finally, there's the emergence of traits that execute truly novel functions.

The fusion of evolutionary and developmental biology – known as 'evo-devo' – studies how animal forms evolved by alterations to the genetic and cellular programs that make them. And just as we discussed the developmental journey of mammalian testicles, the transformation of gonads into testicles by SRY and the making of penises and vaginas, this will remain a recurring theme throughout the book.

Evo-devo frequently answers the questions of *how* an organ evolved. Strictly, it is unconcerned with *why* a structure evolved. The selection pressures that caused one new form to evolve and persist and not another are apart from the processes that generate nature's variety.

One pioneer of evo-devo is Günter Wagner at Yale University. Wagner is especially interested in the distinction between the adaptation of existing structures and innovation – the creation of new structures. Adaptation is concerned with how mammary glands, say, change form according to the specific biology of their possessors, and how, for example, this gland might produce different milk according to the needs of the young. Whereas innovation is concerned with the origins of the mammary gland. No other type of animal has glands that provide their young with sustenance like this.

Generating a mammary gland from an ancestral mammary gland-less state is a very different challenge from adapting a mammary gland once it exists.

At the heart of innovation is the generation of new types of cells. This is achieved by the emergence of new genetic networks, 'character identity networks' as Wagner likes to call them. Wagner distinguishes between the genes that specify a cell's identity and the genes it uses to function. For instance, the genes referred to in Chapter 4 that were switched on in different regions of a developing limb or reproductive tract, to specify the future fate of that region, were parts of such character identity networks. Like SRY, these genes act by switching on and off further identity genes until a pattern of expressed genes creates a cell type that is unique to either an ankle or a uterus, for example.

Once that identity is established a further set of genes – known as 'effector genes' – is switched on that go about the business of making the cell act like a uterine or ankle cell. It's a little bit like a person first assuming the identity of a chef by donning an apron and hat – his or her identity network – but only later becoming a functional chef when provided with knives, pans and a stove – equivalent to effector genes.

'The study of novelties,' Wagner writes, 'is to explain the origin of characters that open up new functional and morphological possibilities to the lineage possessing this character.' He notes that outright novelties occur only rarely.[*] But besides the challenge of accounting for how innovations originate, what is fascinating about that comment is the idea that the emergence of something entirely novel – a mammary gland, say, or hair – creates new possibilities for the organisms that acquire them. As we saw in Chapter 4, internal fertilisation would have evolved for one reason, but in doing so, it allowed for new modes of embryo care and animal development to occur.

---

[*] Indeed, strictly defining when a trait is a genuine novelty versus an extreme adaptation of a pre-existing trait can get technical.

Another example of a novelty that opened new possibilities is a cell type that Wagner, with his colleague Vincent Lynch, has himself studied: the decidual stromal cell, or DSC. The DSC is a cell type that exists only in the wall of the uterus of placental mammals and is fundamental to pregnancy in these animals. Among DSC's various functions, it helps constrain how deeply the placenta embeds in the uterus, and prevents the maternal immune system from attacking the foetus – something that should happen, given that a foetus contains the alien genes of its father. Wagner and Lynch have found that DSCs arose from a cell type, today observable in the uteri of marsupials, by those cells becoming responsive to the hormone progesterone. After some remarkable genetic rearrangements, progesterone now switches on a unique array of genes in DSCs to create a cell that permits a longer and more complex pregnancy than occurs in marsupials. Wagner continues to explore this cell type's origins, and now thinks that DSCs' antecedents had crucial functions in modulating an inflammatory response of the womb to an embryo, to convert that inflammation into the implantation process we see today.

The germline

One of the most fundamental divisions of labour between the cells of an animal is that between those that *make* the animal and those that retain the potential to *make a future* animal, such as sperm and egg cells. Only when life went multicellular did cells come to exist that could no longer give rise to future generations. This notion was formalised in the late nineteenth century by the German biologist August Weismann. In 1881, Weismann began a lecture entitled *The Duration of Life* with a quote from a physiologist of the preceding generation, Johannes Müller, who had said: 'Organic bodies are perishable; while life maintains the appearance of immortality in the constant succession of similar individuals, the individuals themselves pass away.'

Weismann then explained that any cell that contributed to the building of an organism's body was mortal – it died when the animal died. But the cells whose function was to reproduce the organism were essentially immortal – they went on dividing generation after generation.

Weismann expanded this idea into his theory of the germ-plasm. He argued that a lineage of germ cells ran through the succession of animals that were born and died and that these cells were the mediators of inheritance. He thought they were kept entirely separate from all other bodily cells. The emergence of new biological traits could, therefore, only arise from modifications to these germ cells (what we now understand to be changes to the DNA they contain). This insight had a profound and immediate influence, as it negated the old idea that characteristics parents had acquired during their lives could be passed on to offspring.[*]

But this notion of an immortal lineage of cells running through perishable bodies also presented a beguiling concept of what exactly underlies the continuous procession of life.

Most of my life, when I've heard the term germline, I've equated it simply with the sperm and eggs that emerge from testicles and ovaries. But that's not quite right; the germline is a physical reality. There really is a line of cells that runs through the body of every fertile animal that never contributes to the construction of that animal's body.

In mice – the mammal in which this has been best studied – conception is followed by an initial rush of cell generation, then while most of these cells start to move along their mortal coils towards building one or another part of the perishable body, a small pocket of cells is quietly held in reserve. These are the 'primordial germ cells'. Only these cells retain the potential to seed another generation. They are the holders of Weismann's germplasm. After a short stint spent outside of the actual embryo, these cells migrate, via the foetal hindgut, to colonise the gonads where they complete their life cycles. Strictly, ovaries and testes do not make eggs and sperm; rather, they host the germline.

---

[*] Today, there is renewed interest in the possibility that the DNA of gametes is altered in such a way as to change the way it is used by future offspring – a process called epigenetics – but the basic notion that genetic modification is elemental to evolutionary change is secure.

Frankly, what Weismann grasped is that an organism is a means to an end. Our bodies and our minds are aspects of biological structures built to allow *Homo sapiens* germ cells to keep dividing into the future. We are transient vessels in which potentially immortal lines of cells briefly reside.

It seems to me that discovering how unbroken threads of germ cells carry life forward, that is appreciating how germlines make bodies and cast them aside over and over, feels almost as if it should have been – or was – biology's Copernicus moment. Like that astronomer displacing humans from the centre of the universe by seeing that our planet orbited the sun, appreciating the germline takes our everyday notions that we own our sperm or eggs and turns it back to front. We are the germ cells' bodies.

The mammalian way

When those two germ cells met in 1976 to make me, they fused inside my mother. In her fallopian tube they began to divide. The initial structures that the cells made were unique to mammals. A blastocyst formed, where the potential to make my future body laid in a tight ball of cells called the inner cell mass. This mass was surrounded by a larger hollow sphere of cells destined to be part of my placenta – the initial stages of placental mammal development being far more concerned with making this embryonic life-raft than with building the rest of the mammal.

Arriving in the womb, that blastocyst embedded in my mother's uterine wall and placenta-building started apace. Once the placenta was established, my development was fuelled by this structure gathering from my mum everything that embryonic me needed. Placentas are the subject of the next chapter. The key thing is that the mammal initially grows in a womb – I spent nine months *in utero*; an African elephant carries her young this way for nearly two years – the young mammal's world is its mother.

And in mammals, even after the young are born at the end of a long and complex pregnancy, they remain tethered to their mothers by the teat, eating only nutrition that arrives via this elder's body. Twice, I've watched my partner spend a year feeding our daughters this way, for the first six months with nothing solid passing their lips. An orangutan nurses for eight years. And while the suckling youngster remains with its mother – sometimes with a father too – that youngster might also learn essential survival skills, or form the sorts of bonds that define mammal societies. This is mammalian reproduction: the fragility of the developing animal is a parental concern; successive generations overlap in profound and nuanced ways.

Contrast this with our spawning ancestors whose sperm and eggs met in the water outside of their bodies, and who then turned tail and left their developing young to it. These animals – like most spawning animals do today – drew a clear line between generations. In spawning fish, reproduction is a numbers game. A female oceanic sunfish lays up to 300 million eggs in one go. Many, many offspring are conceived, but only a tiny fraction survives.

Even mammals that breed like rabbits, producing what by human standards are many offspring, still invest heavily in each one. Parental, mainly maternal, investment is central to mammalian reproduction.

Sometimes, when Cristina and I look at our daughters, we wonder what they might *become*. We question what these adults will be like, what careers they might pursue, what passions they'll hold, who they will *be*? Always it's a given that a child is a transitional form, a creature growing up towards actualisation.

And, beyond the human realm, this notion feels, perhaps, even stronger. Aesthetically and zoologically, we marvel at adults – at the lion, not the cub; the mare, not the foal; the butterfly, not the pupa. Adults represent evolution's creativity. The young are merely en route to achieving the potential they contain. But this is shortsighted, all life stages must work; to persist a species must make it the whole way around the procreational loop.

What we as parents have invested thus far – Cristina especially – and what remains to be given, represent the way the mammalian parental model has been stretched to extremes in humans. But what becomes clear considering the emergence of parenting, is that we and our daughters are not just successive generations of a species – mammalian evolution yielded *us*. It sculpted a family unit. It honed the interactions between us, so that, for a time, we are each a part of an essential union of two generations.

Placentas and pouches

Few images of mammals evoke maternal intimacy as strongly as that of a kangaroo with her young snug inside her pouch. But this pouch is no simple accessory to carry a young mammal that has developed along lines humans are familiar with. As we consider the early mammalian development and maternal care, we encounter the starkest dichotomy between the two great dynasties of mammals, the placentals and the marsupials.

What have I already said about marsupials? There are about 330 species of them. They live mainly in Australasia but also in South and Central America, and the Virginia opossum has slowly made its way up through North America. Marsupial scrotums lie in front of their penises and these penises are two-headed. The females have twinned vaginas and a separate birth canal. Marsupials have X and Y sex chromosomes, albeit their Y chromosomes are a bit smaller and lack the red-herring gene ZFY. I've said that placental mammals and marsupials diverged about 148 million years ago. Oh, and that in 1770s England, a kangaroo skin became quite a spectacle.

That kangaroo, however, wowed Londoners about 270 years after the first marsupial arrived in Europe. In 1492, the Spanish explorer Vicente Pinzón sailed with Christopher Columbus on his first voyage to the Americas. Seven years later, Pinzón travelled to Brazil to lead the first posse of Europeans to see the Amazon River and, also, an opossum.

The opossum was carrying young in her pouch, and Pinzón was so fascinated by this inbuilt infant carrier he took the animal and her young to present to the court of Spain's Queen Isabella and King Ferdinand.

There are two versions of this event. The one in wider circulation has it that Pinzón carefully transported the mother and her two young across the Atlantic and that Queen Isabella delightedly watched the frolicking youngsters snuggle in and out of the maternal pouch. Then, in a fit of joy, the queen declared that the animal was an incredible mother.

The second one agrees that Pinzón took the opossum to Spain. But in this version, before Pinzón got home, the mother died, and the youngsters were lost. Queen Isabella – instead of watching an irresistible interaction between a mother and her infants – stuck a finger in the pouch and was impressed by the accessory.

You choose. Just bear in mind it was 1500, it took over a month to cross the Atlantic, and opossums don't normally live on ships.

Over the following centuries, European travellers had a steady, if sparse, stream of encounters with marsupials. Owing to how the world's Iberian superpowers divided the globe in the seventeenth century, Spaniards met more American marsupials, while the Portuguese headed south and encountered Australian ones. In 1770, Captain Cook didn't actually discover Australia; he found its eastern coast, allowing map makers to make sense of the area, and also introducing the Old World to Australia's more exotic east-coast natives such as kangaroos and, eventually, platypuses.

It was always the pouches of the newly discovered marsupials that sparked the imagination. The name marsupial is derived from *marsupium* – the Latin for pocket – meaning that, like mammals overall, this group is named after a maternal characteristic.* What was especially provocative was how young the animals in those pouches could be: often the

---

* And marsupials are named for a structure that almost 50 per cent of females don't have, although their youngsters still cling to their abdomens as if they did.

creatures hanging onto their mother's teats for dear life were tiny, furless and positively foetus-like.

Consequently, the common seventeenth-century view was that marsupial embryos grew directly out of the nipples. Willem Piso, a Dutch naturalist, claimed that his dissections had proved marsupials didn't have wombs, and in 1648 wrote, 'The pouch is the uterus of the animal, into this pouch the semen is received and the young formed therein.'

Things only got slightly less fanciful after better dissections found the twinned uteri: it was then postulated that embryos travelled from the womb through the body to be born out of the nipples. This sounds far-fetched, but an alternative (although folksier) idea was that marsupials were conceived in their mothers' noses, then sneezed into the pouch. The mother could often be seen with her head in her pouch just before the young appeared there, and the male's twin-headed penis apparently looked perfectly designed for ejaculating up a pair of nostrils …

To comprehend these bizarre theories, you have to appreciate just how small these pouch occupants are. The largest living marsupial is the red kangaroo, which is about human size, and this animal gives birth to an independent animal resembling the topmost section of your little finger. Actually, the newborn kangaroo is probably a slightly brighter pink, but like your little finger tip, it has no functional eyes or hind legs. As a fraction of its mother's weight, the newborn kangaroo equates to about 0.003 per cent. And in smaller marsupials, the size of their newborns gets even more ridiculous, with a newly emerged honey possum weighing 4 milligrams – 1/250th of a gram (or if you prefer, there are over 7,000 of them to the ounce).

The mystery of young marsupials' arrival in the pouch was only definitively resolved in 1920 when Carl Hartman of the University of Texas finally detailed the birth of one. Hartman captured a pregnant, wild Virginia opossum – an animal that looks a bit like a large white-faced rat dressed in a badger's coat – and kept it in a cage outside a window of his home. Given that it was 1920, Hartman fully and correctly expected the young to emerge from the animal's cloaca, the twinned uteri and transient birth canal now accepted realities. But he wanted to know how exactly the tiny newborns got to the pouch. He and his wife watched the cage night and day.

Finally, Hartman saw that the mother does, indeed, place her snout into her pouch just before the youngsters arrive – but he understood that she wasn't sneezing embryos, just licking the pouch clean. The mother then assumed a sitting position and licked at her genitals. But as Hartman wondered if she might transport her tiny young to the pouch with her tongue, the opossum turned to face away from the window. Quickly, the Hartmans rushed outside, thankfully arriving in time to see the last embryo being born. The genital licking apparently freed the youngsters from the fluid in which they were born.

Hartman then reported that the newborn, looking 'more like a worm than a mammal […] travelled by its own efforts; without any assistance on the mother's part.' To get to the pouch, it crawled 'a full three inches over a difficult terrain', using its precocious forelimbs to essentially swim through its mother's fur. To test the youngsters' instincts, Hartman removed one from the pouch and placed it back at the cloaca to see if it would travel up again to the pouch. It did.

What Hartman describes next gets to the stark reality of marsupial reproduction. The embryo, he says, 'can do more: after it has arrived at the pouch, it is able to find the nipple amid a forest of hair. This it must find – or perish.'

This is true of any marsupial newborn – the nipple is its lifeline, the locus of its next stage of development. But for opossums, as for many marsupials, the search for a teat is a grim reminder that natural selection is an emotionless and pragmatic process, for opossums typically birth more offspring than they have nipples for. Inside the pouch, Hartman saw 'a squirming mass of eighteen red embryos of which twelve were attached, though thirteen might have been accommodated. The remainder were, of course, doomed to starvation.' And he saw how the instincts of the doomed had them hopelessly suckling on a flap of skin or a sibling's tail.

Like their opossum cousins, newborn kangaroos have ratchet-like arms to get them to the pouch. The joeys have further to climb, but unlike opossums, only one kangaroo is born at a time and, at its journey's end, it has four nipples to choose from.[*]

Once attached, all marsupials spend weeks if not months permanently clamped to their mothers' nipples. They may not be born from teats, but most of marsupials' development occurs there.

Marsupials' tiny size as newborns reflects very brief gestation periods: a red kangaroo is pregnant for just 33 days, whereas rabbit-like bandicoots have, at 12 days, the shortest gestation period of any mammal. But long and dynamic lactational processes compensate for limited uterine incubation. The red kangaroo spends six months suckling and developing inside the pouch. Only then does it finally emerge – looking now like a kangaroo – as if born for a second time. It is at this point that marsupials are more developmentally comparable to newborn placental mammals.

Marilyn Renfree, of the University of Melbourne, has said, 'marsupials have, in effect, swapped the umbilical cord for the teat.' Although, it seems possible, too, that placentals swapped the teat for the umbilical cord when the evolution of the decidual stromal cell reconfigured the possibilities of pregnancy.

Other features distinguishing marsupials from placental mammals are more subtle than their disparate reproductive strategies. Purists know various bones and teeth differ in clear ways, which is, of course, essential when it comes to interpreting fossils. Marsupials maintain fractionally lower core temperatures than most placental mammals. They also lack the bundle of nerve fibres that link the two halves of a placental mammal's brain. This latter feature plays into a long-running debate as to whether marsupials are quite as smart as their placental cousins. Marsupial brains might be a touch smaller overall, while the extent and

---

[*] The opossum's 13 nipples also make them the only mammals with an odd number of teats – 12 are in a circle, with one in the middle.

complexity of their social interactions and general behaviour are slightly more limited than those of similar placental species.

In fact, to cut to the chase, marsupials have, for a long time, been regarded as second-class mammals fortunate to have inherited their own island continent on which to experiment. There are, after all, over 5,000 species of placental mammals compared to the 330 marsupials. Placentals span the globe, dominating many ecosystems, whereas marsupials are head honchos only in Australia.

Certainly, in evolutionary biology's early years, when the idea of progress (typically towards humans) was implicit, people viewed placental mammals as higher animals. In fact, when Thomas Huxley renamed the major groups of mammals in 1880, he termed placental mammals 'Eutherians' or *true* mammals, monotremes 'Prototherians' or *first* mammals, and marsupials were termed 'Metatherians' – essentially, *halfway* mammals.

Marsupials are not halfway to anything: they are their own adaptive radiation. But is this inferior label fair? The point is moot. Those who say yes, emphasise that marsupials once lived worldwide but ultimately lost out to eutherians on all continents bar Australia. The alternative view is that marsupials are simply different and that they – their reproductive biology, especially – are better adapted for life in grimmer, more erratic conditions. The outback is a tough place to live. And the marsupial mode of reproduction may give mothers greater control over their investment in their young, including being able to terminate pregnancies or lactation more easily if resources become too scarce.

But talk of differences distracts from what is most striking about marsupials and placental mammals, which is how similar they can be. As European explorers kept finding more and more marsupials, it became the norm to name them after what was deemed their Old World equivalent. Often the English translation of a marsupial scientific name yields a prefixed familiar one: pouched dog, pouched badger, pouched hare, bear or antelope, and so on.

And while some of these names feel somewhat lazy – a koala *bear*? – it's true that many pairs of marsupial and placental mammals are arrestingly alike. The marsupial mole and Africa's golden moles both swim through sand, leaving no tunnel as the sand collapses behind them. The flying marsupial phalangers glide between trees in much the same way a placental flying squirrel does. Wombats look like groundhogs. 'Marsupial mice' really are very mouse-like. And the rabbit-eared bandicoot doesn't just have rabbit ears, it has powerful, hopping rear legs too. The *thylacine*, which went extinct on 7 September 1936 when the last known one died in captivity, was also called the Tasmanian wolf and was strikingly wolverine. Until 50,000 years ago, there were 'marsupial lions'.

Even when the resemblance isn't immediately obvious, a marsupial and a placental mammal often play similar roles in their respective ecosystems. Numbats look sprightlier than placental anteaters, but they hoover up the same diet. And despite its rather different anatomy, a kangaroo functions in the outback much as antelope, deer or other large grazers do elsewhere – both these groups even independently evolved foregut fermentation to digest their food.*

When biologists talk about a 'last common ancestor of marsupials and placentals', that species can feel like a hypothetical conceit of an abstract theory, no more than a branch point on an inferred phylogeny. But to accept the theory is to accept the reality of that animal. From that last shared ancestor, two species emerged that would have been

---

* There are a few notable gaps in the marsupial repertoire, most notably an absence of both bat-equivalents and aquatic forms: no marsupial has truly taken to either the sky or the water. The most plausible explanation is that marsupial reproduction places restrictions on evolvable forms. First, an animal that needs to suckle constantly on the underside of its mother is unlikely to thrive underwater. And second, the need to ratchet one's way from cloaca to teat as an embryonic newborn may require a forelimb design that negates the later development of wings and fins.

barely distinguishable from each other, but the pair of them were the starting points for two full-blown radiations of different types of mammals.

That eventually these separate radiations produced so many convergent forms – *entirely independently* – is phenomenal. First, it's testimony to how powerfully natural selection melds organisms to exploit particular ecological niches. And, second, it seems to emphasise further how similar marsupials and placental mammals really are – how the fundamentals of mammalian biology were in place before their great split about 148 million years ago, and how the potential lying within each lineage overlapped to a huge degree.

Thirty-four years after that fateful fusion of egg and sperm began the construction of me, I let the germline continue. Poor old germline, all those millions and millions of years of surviving, evolving, winding its way to me and what does it get in return? Years of latex barriers and encounters with women who, via the pills they took, didn't even let *their* germ cells escape their gonads … Anyhow, we got there. Another procreational loop.

Writing about marsupials recalls how in the neonatal unit when we took Isabella from her crib and held her on our chests, they called it 'Kangaroo Care'. At the time, I assumed the name referred to the pouch-like arrangement of blankets that enwrapped our daughter as she cosied into us. It turns out, the Colombian doctor who first advocated this type of care was inspired by the immaturity of a newborn kangaroo and the alternative developmental course it took.

Afterbirth Before Birth

After Cristina's waters had broken two months before Isabella was due, we processed everything that happened – the cab ride, her admission to hospital, all the initial consultations – in streams of cognition that were purely sensory. Our brains had raised barricades around their emotional centres. We only began to accept our reality once we were in a private room phoning friends and family to explain what was happening.

It had been a fretful, sometimes frightening, pregnancy. Twice, we'd rushed to the hospital because Cristina's womb was contracting aggressively. Each time, she'd been monitored and hydrated, then discharged without a confident prognosis. Told now that she would be hospitalised until the baby was born, we were in the strange position of wishing for Cristina to stay here for weeks.

Instead, over the mere 36 hours between our arrival at the hospital and the moment of imminent birth, we experienced a succession of nursing and doctoring hits and misses. Ultimately, though, our luck seemed good – Cristina's usual obstetrician was on duty, and we were in the care of a nurse who was calm and competent. When the doctor announced that things were about to step up a gear, the nurse smiled and said, 'See you in a sec.'

I went to the bathroom, rolled my shoulders, took a deep breath, and pleaded into the mirror to find some semblance of calm and strength. We still had no idea how our daughter would emerge, but I was not the person who was about to give birth. I was the person who needed to hold his shit together, no matter how scared he was, to be of some use to the person who *was* about to give birth.

Then, when I emerged, our nurse returned wearing a plastic apron, latex gloves, a face mask and galoshes.

'Oh,' Cristina and I said in unison.

She insisted that the storm gear was standard, but the damage was done. *What on earth*, we wondered, *is about to happen?*

What did happen over the next hour, or however long it was, was a whirl of amazingness. Our doctor, in full control, directed everything, but subtly so; she managed to make the entire event about us. And then we had Isabella, and I wept – good, deep happy sobs – and although she weighed only 1.8kg (4lb) she cried out, and we knew despite her being whisked off to another ward that she was OK.

I simply can't imagine that I would have been anywhere near as moved if Cristina had laid an egg. Or maybe I would have been. Maybe it would have been a jubilant double whammy? First, we'd have celebrated the laying, and then we'd have got really choked up at the hatching. Who knows?

In all the emotion, I paid very little attention to the rather sizeable clean-up procedure that happened around us. Except I did glance at a plastic tray that our nurse was carrying off, briefly contemplating its mass of purple-red contents.

Later, I tried to hug and thank the obstetrician for delivering our baby. She ducked out of this without even pausing her conversation about the next expectant mother she'd attend. But really it had been that plastic tray's spent purple mound that had delivered our infant to this point. That particular placenta was on its way to a pathologist to try to understand why the pregnancy had ended prematurely, but usually this most remarkable of organs, its job done, is simply thrown away. Or, by many mammals, eaten.

When our second daughter, Mariana, was born, I was equally elated and exhausted, but much less confused. She was born pretty much at full term, and the post-partum atmosphere was decidedly less fraught. What a marvel it had been to see her immediately seek and find her mother's nipple. Also, I had by this time a strong suspicion that I was going to write a book about key mammalian attributes, so I made a point of taking a decent look at the afterbirth. I peered into another plastic tray where the placenta lay, to try and take in this transient organ before it was dispatched to the incinerator.

It was a very unfamiliar purple; it had almost a lilac sheen. There were a few prominent blood vessels coursing through it, their meanderings like rivers forking and winding across a landscape. The smaller kinks appeared random, but in the larger scheme, their trajectories looked purposeful. Otherwise, it was amorphous. Nothing about it indicated its wondrous function.

For many biological entities – the heart, say, or the hand – you can appreciate how their structures define their functions; how they operate is inherent in their observable form. This notion that structure and function are inseparable is true of all biology, but for many things the relationship is only apparent if the tissue can be interrogated microscopically. I once held a human brain in my hands, and wrestled with how that inert, almost formless lump of flesh had once created a person who had danced and suffered and loved, and who had held thoughts and ideas. This placenta reminded me of that – it was unfathomable. I stared at it only long enough to know I was grateful to it and long enough to know that no spark of understanding was going to leap from it. Besides, another mammal had just entered into this world, and she and I needed to hang out.

Since that night, I have discovered that, as I stood gawking into that plastic tray, I knew even less about placentas than I thought. In the context of this book, my most notable mistake was thinking that organ was something uniquely mammalian. The clue seemed to lie in commonly used terminology – after all, the three people in that room were *placental mammals*. This being a name born of (1) the plastic tray contents and (2) the gland on which my daughter was suckling, I assumed each characteristic was equally definitional of placental mammals. Additionally, I'd assumed that my daughter's emergence unenclosed in eggshell was also a rare phenomenon – another mammalian quirk in a world full of egg-layers.

In reality, however, while mammary glands are indeed very special, placentas and live birth are less so. Live birth – or 'viviparity' – has evolved in vertebrates alone about 150 times; most notably, various lizard and snake lineages have independently concocted it about 115 times,[*] though birds, crocodiles and turtles never have. Live young emerge from a small minority of amphibians and bony fish and from many cartilaginous fishes, particularly sharks. Youngsters getting a developmental head start inside their mothers is something evolution has frequently favoured. John Ray's 1692 proposal to name mammals *Vivipara* was a terrible idea.

How about placentas? Are they more special to mammals? Well, a little bit, at least. Among viviparous species, an important distinction is made regarding how development is fuelled. It can either be powered solely by resources the mother packaged into her eggs – as is the case in the simplest type of viviparity, where eggs are retained in the oviduct until they have hatched – or a youngster can feed off other maternal resources as it grows. Such feeding, termed 'matrotrophy', can be achieved in various ways: some embryos absorb goodies released by specialised maternal glands, while some eat in a conventional way, ingesting the uterus lining, or unfertilised eggs that the mother ovulates, or their siblings! Sharks do the latter until only a single victorious young shark emerges. A placenta is the final mode of feeding, this organ being a specific part of the growing embryo that interfaces with the maternal reproductive tract, so as to take maternal resources that the embryo uses to grow.

In vertebrates, matrotrophy has evolved a more modest 33 times, and placentas have developed on maybe 20 occasions.

What then makes mammals special? Well, first, viviparity was a feature of the last common ancestor of marsupials and

---

[*] Chapter 2's description of the quest to determine if platypuses laid eggs may have seemed quaintly archaic, but still today, naturalists strive to actually witness certain wild lizards either laying an egg or birthing live young.

placental mammals,* so all therian mammals share the trait. Second, although many offspring consume maternal resources, most initially use resources from an egg yolk, whereas therian mammals' egg cells are greatly diminished – pretty much everything that goes into making a therian mammal arrives via a placenta or teat. Finally, in mammals, placentas reach unparalleled sophistication.

The mammalian placenta is an organ made of foetal tissue enmeshed with maternal uterine tissue through which foetal blood vessels run, linked to the foetus via the umbilical cord, to maintain and nourish a growing mammal for the first stage of its life.

The other major point of my naivety in the delivery room was in imagining that the placenta I beheld was a straightforward emblem of maternal benevolence.

Structure

In the winter of 1750, a recently dead young woman was heaved through the back door of an anatomy school in Covent Garden. Nothing unusual there; John Hunter, the younger brother of the school's proprietor William Hunter, claimed that in 12 years working there he saw 2,000 human corpses dissected. Most of them were lifted from graves and brought to the back door by John's underworld associates. But this arrival was special – William was thrilled – for the body contained a second corpse; the young woman had been nine months pregnant.

John and William Hunter were a fascinating double act. William was the seventh born and John, a decade later, the tenth and last child of their Lanarkshire farmer parents. Only two of their siblings also reached adulthood. The parents gave William every educational opportunity they could: at 14, he read divinity at Glasgow University, but then decided medicine was his calling, and so his parents arranged training

---

* Although it has been suggested it evolved early and independently at the roots of each of these radiations.

from two prominent Scottish physicians. After he had sailed to London at the age of 23, he studied under another two esteemed doctors, both experts in anatomy and midwifery. In the capital, William 'favoured the wearing of a rather full wig'. He was eloquent, entrepreneurial and aspirational.

John was famous for his gruff ways. As a child, he never liked book learning. He often bunked off school to explore the local forests, and after William's departure, John dropped out of school altogether to care for his widowed mother. But then, at 17, he bade her farewell and went to London to find his brother.

When John arrived, William presented him with a human arm to dissect. It's uncertain whether William knew John had spent years taking apart the corpses of Lanarkshire forest animals, but John was a natural with the knife. Impressed, William gave him a second arm. This one had coloured wax injected into its blood vessels. John's task was to cut away the flesh leaving only the wax casts. Again, he executed the task with aplomb. John and William were perfectly complementary. The elder lived for his onstage lecturing, while John was at his best with his hands inside a cadaver, exploring, discovering; flesh under his nails.

When the corpse of the pregnant woman arrived in 1750, John was immediately put to work dissecting her. With him was Jan van Rymsdyk, a Dutch artist William had hired to draw what was revealed each time John peeled back another layer. So little was known of human pregnancy that there was much to be gained from offering new insights into it.

The most famous drawing of the nine-month-old foetus is startling. Today, ultrasound scans are so ubiquitous, we believe ourselves to be familiar with the inside of a womb. But ultrasound produces only fuzzy caricatures – its images may make parents' hearts race, but only because of what they represent. Van Rymsdyk's drawing appears photographic. At first glance it is beautiful. The baby looks full of life – only shy, its face turned away. It is about to engage. It is about to enter the world. Only it isn't. And once that fact registers, the image's beauty collapses into repulsiveness.

The Hunters, however, were anything but repulsed. Between 1750 and 1754, they acquired the bodies of five pregnant women and one of a woman who'd recently given birth. John dissected, van Rymsdyk drew, and William plotted what would be hailed as his masterpiece. *The Anatomy of the Human Gravid Uterus*, published in 1774, was a spectacular portfolio of illustrations working back from the initial full-term child to a three-month-old foetus.

Besides offering the world this glimpse of how a human foetus grows, the Hunters provided two important insights into human pregnancy and the placenta. First, while John dissected the original case, injecting coloured wax into the womb's blood vessels, William described the thickened uterine lining that was the maternal part of the placenta, the tissue that would normally be shed at the baby's birth. He called it the 'decidua'.

The second, more profound discovery was made in 1754 after John Hunter was summoned from Covent Garden by Colin Mackenzie, who was also trying to make sense of pregnancy. Having injected yellow and red waxes into a foetus and uterus, Mackenzie ran to fetch him. Hunter was enthralled. The injections revealed what his previous attempts had failed to. They confirmed that the two circulatory systems were entirely separate; foetal and maternal blood never mixed.

John Hunter later wrote that William initially failed to grasp this observation's importance – that he didn't see that it quashed old ideas that maternal blood circulated through the foetus to sustain it, thus demonstrating a fundamental independence of the baby.

However, once William *did* understand, he lectured extensively on the finding and included it in *The Anatomy of the Human Gravid Uterus*, a book that acknowledged John Hunter merely for his assistance and didn't even mention Jan van Rymsdyk. When it was published, the Hunters were both Fellows of the Royal Society. William was a professor of anatomy and physician to the queen. John was a renowned surgeon, and his commitment to using observation and basic anatomy to improve surgery had begun to transform surgery from the archaic and barbaric pursuit he'd first encountered into a new discipline. His medical career had begun when William paid for him to attend medical school. But as William ascended through London society, John remained focused on his scientific interests. He worked on surgical technique, experimented with transplants and, whenever he could, investigated curious zoological phenomena.[*]

John even became convinced that some form of evolution had occurred over Earth's history. At one point, as tensions grew between the brothers, students could, in the same week, attend lectures in which John speculated about evolution and listen to William celebrate the human body as evidence of God's omnipotence.

No one, though, anticipated that on 27 January 1780, at a routine meeting of the Royal Society, John Hunter would take the stage – ostensibly to present his paper 'On the structure of the placenta' – and use the platform to call his brother a plagiarist. John claimed full priority for the discoveries he'd made 26 years previously about placental blood flow and decried his brother's dishonesty. William would write once in protest to the Society, John would once reply, but after that day – despite John tending to William on his deathbed – their relationship was over. After all they'd done to help each other, to drive each other forward, the rivalry and the conflicts meant they were forever estranged. Families create the most complex of dynamics.[†]

---

[*] For instance, the hermaphroditic sexual anatomy of freemartins that would inspire Alfred Jost to investigate the hormonal basis of sex determination was first described by John Hunter.
[†] This story has two ironic footnotes. First, unknown to the Hunters, a Dutchman named Wilhelm Noortwyck had shown that foetal and maternal blood systems were separate a decade before Hunter and Mackenzie did, albeit Hunter's evidence was perhaps stronger. And second, if the way William took credit for his brother's work was distasteful, John's brilliance would be even more abhorrently

No one knows if John Hunter discussed his ideas about evolution with Erasmus Darwin. Darwin is, today, unjustly best known for having had a famous grandson. But he was a brilliant physician and scientist in his own right, who in 1753 attended classes at Covent Garden and apparently got on very well with Hunter. Hunter would doubtless have approved of his former student's 1794 book *Zoonomia*, in which Darwin argued that life was sculpted by evolutionary processes. These prescient ideas are the second thing that Erasmus Darwin is famous for, but, in *Zoonomia* he also provided an essential insight into the placenta.

With Hunter's conclusive demonstration that maternal and foetal blood systems were separate, the placenta had been acknowledged as an interface of some sort, an organ that mediated exchange between mother and baby. Darwin was fascinated by oxygen, which had been discovered in the 1780s. As this gas's intimate ties to life were rapidly explored, blood was found to change colour from dark burgundy to bright red upon exposure to it – a colour shift that happened when blood passed through either lungs or gills. Erasmus Darwin reported that it also occurred when blood moved through a placenta. He wrote that the placenta 'appears to be a respiratory organ like the gills of a fish by which the blood in the foetus becomes oxygenated'.

It was an observation that resolved the longstanding puzzle of why foetuses *in utero* don't suffocate. However, *Zoonomia*

---

plagiarised by another family member after his death. This thief is a familiar character in this book. In 1771, John married the poet Anne Home, whose younger brother was named Everard. The dissection skills used to investigate platypuses, echidnas and kangaroos when they first came to Europe were taught to Everard Home by John Hunter. However, while Home was no doubt talented, he was not the sole source of his apparently huge productivity and wide breadth of insights. After his brother-in-law's death, Home stole John Hunter's unpublished manuscripts and published their contents as his own. And when suspicions intensified, he burnt the originals to try to hide his crimes.

also propagated an old untruth. It asserted that the foetus grew by ingesting sustenance from the amniotic fluid. This fluid was imagined to be like the white of a bird's egg – similarly, it was initially abundant, then diminished as the infant grew. In 1794, with no clue about molecules and their trafficking between cells, it was a fair mistake to make.

Erasmus's grandson paid curiously little attention to placentas. Despite his staggering breadth of interests and his conviction that embryology held fundamental clues to evolution, Charles Darwin never truly discussed the organ. However, placentas featured prominently in one of the bitterest debates to erupt after the publication of *On the Origin of Species*.

Richard Owen – that staunch opponent of evolution and platypus egg-laying – believed mammals should be classified according to whether or not their brains' surfaces were convoluted. Using brain anatomy also allowed Owen to produce a third group, discontinuous from all other mammals: only humans, Owen said, had a hippocampus minor in their brains.

Darwin's self-anointed 'bulldog', Thomas Huxley, said giving the brain surface priority resulted in ridiculous groupings of mammals – obviously disparate groups were brought together and closely related forms were divided. He then triumphantly dissected a hippocampus minor from an orangutan brain.

Huxley's alternative, proposed in 1864, was a mammalian taxonomy based on evolutionary principles, and he said that the best trait from which to derive genealogical relationships was the placenta. In this, he was following previous workers who'd stated that the absence of placentas was what distinguished monotremes and marsupials from other mammals, the latter becoming known as members of *Placentalia*.[*]

---

[*] This whole episode illustrated rather well a remark made in the 1930s that the first attempts to classify organisms according to their genealogical relationships, as per Darwin, produced groupings that weren't much different in type from those previously produced by non-evolutionary approaches. The groupings could be different, as

To divide placental mammals, Huxley looked to an 1828 essay that described four different shapes of placentas and paid particular attention to how extensively uterine tissue contributed to the placenta. Such maternal elements were William Hunter's decidua, the part of the uterus that expands to interact with the foetal component. Two shapes of placenta were apparently associated with a pronounced decidua that was bloodily shed at birth, whereas two placenta types had much lesser uterine components. Using these shapes, Huxley found the resultant groupings of placental mammals much more pleasing than Owen's.

The satisfaction was, however, short-lived. A healthy amount of variability in a trait across different species is useful for inferring those species' relationships. But placental structure, it turned out, was astonishingly variable – it's now understood to be Mammalia's most changeable organ.

Curiously, although this was apparent within decades of Huxley's original proposal, people fruitlessly tried to infer mammalian relationships using placentas for another 50 or 60 years, and then sporadically for another century. Why this was a fool's errand and why placental types were so unpredictable would only make sense once a better understanding of placental function had been achieved.

The arrival of evolutionary theory was doubtless the landmark event in nineteenth-century biology, but a second – equally important – revolution occurred in parallel: cell theory was born, whereby biologists got wise to the fact that all organisms were made of cells. And from appreciating this, it became increasingly clear that understanding cells' forms and functions was necessary for comprehending both organs and organisms.

The growth of cell biology was driven by the ever-increasing power of microscopes (plus improvements in tissue preparation and staining). As anatomists explored the possibilities of these devices they saw ever more clearly the

---

Owen's and Huxley's were, but in both cases, similarities between traits were their basis, just as they were in Linnaean taxonomies.

fine-scale structure of living tissues. For every organ, this was rewarding. But for structures such as brains and placentas, where gross morphology told so little of their machinations, it was transformative.

Trained on placentas, microscopes revealed meshes of blood vessels interwoven with ambiguous layers of other tissues and curious spaces. The challenge became understanding which parts of this morass were foetal and which were maternal. For decades, this was almost guesswork.

One of the most crucial breakthroughs came at the end of the nineteenth century when the Dutch biologist Ambrosius Hubrecht was inspired by Huxley's claim that insectivores – shrews, moles and hedgehogs – most resembled the original mammals to investigate insectivore placentas. Hubrecht thought they might illuminate what the very first mammalian placentas had been like. His seminal paper studied hedgehogs, which he caught from the wild during their breeding season.

Hubrecht's great contribution was defining the 'trophoblast'. 'Blast' describes an embryonic cell, and 'tropho-' means 'to nourish'. Hubrecht saw that the cells on the leading edge of the expanding placenta didn't just abut the uterus wall, they ate into it. He saw, too, that placental tissue surrounded pools of maternal blood in the decidual tissue of the womb, 'utilising it most fruitfully'. Trophoblasts were the means by which embryos fed themselves, taking nutrients – as well as oxygen – from the mother's blood.

Today, we know that the trophoblast is the elemental cell-type of the placenta. Once the early embryo embeds in the uterine wall, trophoblasts start to digest the maternal tissue, basically burrowing into the womb. This expanding envelope of tissue around the embryo pushes forward towards maternal blood vessels, which themselves are transformed to pool blood for the embryo's benefit.

With some basic understanding of what a placenta does, we can return to the variety seen across mammalian placentas. Detailed analysis of them in different species has shown that in addition to placentas differing in their large-scale structure, they also diverge according to (1) whether one, two or three layers of maternal tissue separates foetal blood from maternal blood and (2) which of three distinct forms of extensions the foetal tissue sends into the womb. A simple idea would be that placentas progressively evolved so that fewer layers of tissue separated the foetus from the maternal blood, and that they produced ever more elaborate extensions. With human placentas having the most complex type of extensions and very little tissue separating the foetal and maternal blood, this idea chimed nicely with Victorian notions of evolution having produced animals that had ascended ever closer to human perfection. But then, nice ideas can be wrong.

The original trophoblast paper ended with Hubrecht calling for Huxley's placenta-based classification scheme to be rejected. Comparing the placentas of hedgehogs to those of moles and shrews, he saw that despite the obviously close relationships of these three insectivores, their placentas were starkly different. Plus, like human ones, they were highly invasive, hardly, it seemed, befitting an early form of placenta.

Hubrecht suggested that perhaps the newness of the mammalian placenta, 'the youngest organ in the mammalian economy', meant that natural selection hadn't had time to achieve its 'pitiless elimination […] of certain placental adaptations which, in the long run, must prove less favourable than others'. The problem here was that Hubrecht mistakenly assumed that natural selection would act on the placenta as it would on any other organ.

Function

The desks in Ian West's classroom formed three sides of a square, and in the centre of the open fourth side, Ian lectured while writing on an overhead projector. Each class he'd introduce a subject, present some necessary background, then, at a pivotal juncture, pause to fire out a question. Always, he demanded that one of us – his A-level biology students – made the crucial conceptual leap forward.

Ian never taught us about placentas. But, every time I think about the forces that shaped their evolution, I think how he would have loved teaching us about them. I imagine him first asking, 'So what's the most important difference between a fish spawning and mammalian reproduction?'

Everyone would have known that he wasn't asking us to state the obvious, like telling him fish laid eggs and mammals birthed freely moving young. Silently, we'd have all fought to come up with something intelligent to say, until Ian broke the silence by barking a random name.

'Umm … fish lay millions of eggs and mammals produce just a few offspring?'

'Yes …' he'd have said in a way that indicated the answer lacked sufficient depth. Then another name.

'Fish eggs are fertilised externally?'

Another, more emphatic '*Yes* …' This one signalling that we should run with something implied by that statement.

What he would have wanted – what he would have coaxed us towards – was for someone to offer, 'Is it that in fish the mother dedicates all her resources to her eggs *before* they're fertilised, but in mammals the mother provides nearly all the resources *after* fertilisation?'

'Yes!' Ian would then have exclaimed. 'Now you're thinking like a biologist!'

A female fish that breeds by spawning fixes her investment in the next generation when she makes her eggs. She partitions to those ova all the energy she is wagering on the propagation of the genes she has also packaged into them. Once at sea, those eggs can ask no more of their mother's physical resources. And, crucially, paternal genes that enter an egg to co-direct the development of the embryo also have no access to the mother's reserves of energy.[*]

---

[*] I am aware that my spawning fish are frequently a little cartoonish; certain fish do, to some degree, parent. For example, some guard their broods, a small number even provide food to their young,

In this regard, egg-laying reptiles and birds are more like fish than mammals. Consider the chicken. One, because the chicken egg nicely represents the basic amniote egg. Two, because a chicken egg is familiar. And three, because I want to say 'This is a chicken-and-egg situation' and mean it not as a tired – if useful – metaphor, but literally.

Although it's always struck me as odd that chickens lay quite considerable eggs even when they haven't had sex, this is standard amniote operating procedure. All females release unfertilised eggs – humans, once a month.* It just happens that chickens have been bred to ovulate frequently and they're housed apart from males.† Plus, mammal eggs have become so diminutive, they're barely noticeable.

Also like mammal eggs, the chicken egg is fertilised high up in the reproductive tract, before tumbling down the oviduct towards its next developmental stage. But unlike mammals – and very much like fish – chickens ovulate eggs with large yolks replete with nearly all the resources needed to make a chick. Then, fertilised or not, the eggs are provided with egg-white – a few final resources – then encased in shell. So, again, all that's needed to make the future hatchling is provided before a single paternal gene is switched on in the embryo.

By contrast, in the womb-burrowing, maternal-investment-grabbing mammalian placenta, paternal genes are active from the start. Because of this, regardless of what the father himself does post-copulation, his genes can – and do – have profound effects on how his offspring develop.

---

while, as noted earlier, some have evolved internal fertilisation and rudimentary placentation.

* Menstruation – the shedding of a portion of the uterine wall with an unfertilised egg – is a relatively uncommon phenomenon seen in humans, other primates, certain bats and the elephant shrew. It reflects a particularity of how wombs in these species shed the linings that have been prepared for implantation. Other mammals reabsorb the lining.

† The success of this strategy is half the reason there are about 20 billion chickens on this planet, far more than any other amniote.

The seeds of understanding the significance of this were sown in 1974 when Robert Trivers at Harvard published a paper succinctly entitled 'Parent–offspring conflict'. In it, Trivers attempted to formalise exactly what characterises the relationship between a parent and its offspring.*

A parent – as anyone with kids knows – only has so much time and energy, and well before Trivers' work, the best way a parent could allocate his or her resources to maximise reproductive success was an evolutionary conundrum. People agreed that natural selection would shape how parental resources were allocated and that this would affect how many offspring parents had, how big they made these offspring and how much energy parents reserved for staying alive and possibly having more offspring in the future. Such theorising successfully distinguished between strategies of producing many low-cost, low-survival young (as fish do) and of making a few heirs that are heavily invested in and much more likely to survive (as we mammals do).†

Trivers, however, said that previous studies had all missed something essential: older work had wrongly viewed offspring as *passive* recipients of parental investment.

'Once one imagines offspring as *actors* in this interaction,' Trivers wrote, 'then conflict must be assumed to lie at the heart of sexual reproduction itself – an offspring attempting from the very beginning to maximize its reproductive success would presumably want more investment than the parent is selected to give.'

Imagine a mother with triplets. And picture that she has just enough resources – food, milk, money, whatever – to get those three youngsters to the stage where they can go off and fend for themselves. *Her* reproductive success would be best

---

* Originally the paper was to be called 'Mother–offspring conflict' as Trivers was mainly focused on mammalian mothers, but he eventually felt the principles extended more broadly.
† Albeit, similar arguments do still apply across mammals too, accounting for the large litters of rabbits and rodents, say, and the single offspring of orangutans and elephants.

served by dividing her resources equally between her offspring, so that they matured equally and all might provide her with grandchildren. The old models, where the triplets were passive, predicted that exactly this would happen.

Trivers, though, pointed to a reality in which each triplet hustled to get more than an equal share of the mother's resources. In fact, they probably competed to obtain *more* resources than the mother would optimally give, diminishing her chances of breeding again. And sex was to blame for this.

Say that the three triplets had different fathers.* If the father of one of the triplets transmitted paternal genes that made this daughter a better hustler – better at getting her mother's attention, say, or stronger at suckling – then she would get a greater share of maternal resources and be more likely to thrive at the expense of her two siblings. Here the daughter is in conflict with her siblings – the offspring in competition with each other.

Additionally, though, if this hustler daughter successfully grabs more resources than her mother would otherwise have given, then she's increased her father's success at the expense of her mother. Daughter and mother, by way of paternal genes, are in conflict. And this can happen even if there's only a single offspring; that offspring's genome being different from its mother's always creates the potential for discord.

Trivers' paper was no saccharine tale of harmonious family life. But his case was convincing – a father who passed on infant-hustler genes was going to sire more offspring that were likely to thrive and to give *him* more grandchildren.

The inspiration for this work apparently came from Trivers watching juvenile monkeys plead with their mothers for more milk when the mother had decided it was time they were weaned. However, conflict of this nature was later understood to express itself most fully earlier than this, inside the mammalian womb.

---

* This can happen when females mate with multiple males in a short period of time. And the same arguments would apply to three offspring born in successive rounds of breeding.

If you take turtles – as Fredric Janzen and Daniel Warner, of Iowa State University, did in a 2009 study – you can calculate what, for a given amount of investment, a mother's best reproductive strategy would be regarding the number and size of eggs she should lay. She could lay lots of little eggs, an intermediate number of medium ones, or a few large ones. The bigger the egg, the more likely the hatchling is to survive, but the fewer she can lay. Janzen and Warner's calculation said the turtle would be best served by laying a medium number of medium eggs. And when they looked at what she actually did, they saw she did exactly what they'd predicted. That is because, like chickens, turtle mums are in sole control of what resources are packaged into each egg – their offspring are passive, there is no conflict.

Now let's consider mammals. In a series of papers beginning in the late 1980s, David Haig, another Harvard biologist, explored the numerous possible arenas for mother–offspring conflict during mammalian pregnancy and came up with quite a list.* From the moment a mammalian embryo implants into the uterine wall, paternal genes are able to make that embryo put its own interests before those of its mother. Passive, a placenta is not.

Haig discussed how at the very beginning of pregnancy trophoblasts release enzymes that digest maternal tissue and maternal tissue releases chemicals that inhibit those enzymes. Additionally, the placenta releases a growth factor that accelerates its own development, and maternal decidual stromal cells release proteins that neutralise this growth factor. From the very start, there are placental actions and maternal counteractions. Haig sees this dynamic interplay as the result of a process akin to an arms race between predators and prey, or between hosts and parasites: one player tries to get the better of the other, and the other evolves to neutralise that threat.

---

* If the idea that a placenta was a uniquely mammalian entity wasn't sufficiently diminished earlier, know that Haig drew much early inspiration from studying the plant equivalent of the placenta.

Another thing the placenta does is release hormones into the maternal bloodstream that manipulate maternal physiology. Hormones are what make the mother's body 'know' it's pregnant, adapting its physiology to the residence of the embryo, then foetus, but this endocrine channel is another means for the young to manipulate its elder. For example, the placenta releases one hormone that causes maternal insulin resistance, thus increasing blood glucose. Yet, as is the way of things, in response to this hormonal hijacking, mothers have again counter-evolved to dampen the placenta's effects; they make more insulin, and maternal receptors for various hormones have also adapted.

Following this line of reasoning, it became apparent that certain genes functioned to accelerate foetal growth while others – some in maternal tissue, some in the foetus itself – worked to slow foetal development. From this came a hypothesis regarding why some genes are switched on in offspring according to whether they were inherited from mum or dad. If a gene speeds foetal growth, the mother might do well to switch off the copy of it that she passes on to her offspring. If the gene suppresses growth, by whatever means, the father will benefit from turning it off in his sperm. 'History matters,' says Haig; the sex of the parent in which DNA is prepared for child-making determines which genes leave the parental body switched on or off. The foetus ends up, says Haig, with a 'paternal foot on the accelerator and a maternal foot on the brake'.

This process is known as genetic imprinting. It is seen in both marsupials and placental mammals, but is more prevalent in the latter – over 200 genes are imprinted in humans – and now it's appreciated that genes imprinted by their different parents continue to influence young mammals after they leave the womb. One 2018 study even showed that imprinted genes alter the brains of mother mice to influence maternal behaviour once youngsters were born.

Unlike the control a female turtle has over her eggs, a mammalian mother is engaged, from almost the moment she conceives, in a dynamic relationship with her offspring and the paternally inherited genes they contain.

Haig's theory of paternally and maternally inherited genes pulling the placenta in different directions helps explain why the organ is so varied across mammals. Paternal genes yank it one way, maternal genes push it another, and so, over history, the placenta frequently changes form. These shifts – often dramatic and occurring between closely related groups – are why placentas are not very useful for inferring relationships between mammalian lineages.

Only after the mammalian family tree had been resolved by genetic means (see Chapter 9) could researchers go back and transpose data about placental form onto that tree to see how this organ has evolved and what it might have looked like in the first eutherians. Two groups have attempted this. They agree that the placenta – with respect to the amount of tissue separating foetal and maternal blood, the shape of its extensions and its large-scale shape – has changed drastically on numerous occasions. However, one study concluded that the original eutherian placentas were of the most invasive kind, whereas the other deduced that they invaded only to an intermediate degree. Given the dynamics of this organ and the conflicts it harbours, there may always be a basic uncertainty here.

To go further back, and ask where placentas came from in the first place, we welcome back – naturally – monotremes. While the discovery that platypuses and echidnas laid eggs was revelatory, monotreme eggs are, in fact, fundamentally different from other amniote eggs. The 4mm-wide (0.16in) eggs that a platypus ovulates are big by mammalian standards, but the eggs a platypus *lays* – about 17 days later, a considerable interval for an egg-layer – are even bigger: they're about 16mm (0.63in) long and 14mm (0.55in) wide. Which is to say, platypus eggs grow substantially in the maternal reproductive tract.

In a platypus egg, the membrane around the yolk (as well as other embryonic membranes) assimilates materials that the mother secretes into her uterus. This gathering of nutrients post-fertilisation says something very interesting. Viviparity and matrotrophy normally evolve by eggs starting to hatch *in utero*, and the hatchlings only then figuring out ways of getting food there. The platypus says mammalian ancestors evolved matrotrophy before they evolved live birth.[*]

It's also worth noting another prominent feature of monotreme reproduction: their hatchlings are helpless and dependent on maternal care. This is probably the ancestral state for mammals. The advent of mammalian viviparity likely occurred in diminutive shrew-like species that were so small that laying viable eggs through their tiny pelvises became impractical. An initial form of internal hatching would have allowed small young to exit and then be sustained by their mother's milk, so this scenario was probably made possible by the prior evolution of lactation.

Turning to marsupials, although the way these animals breed today contains many specialised features that evolved specifically in the marsupial lineage, it is likely that, in some ways, they resemble an intermediate stage in the evolution of placental mammal reproduction. After fertilisation, the initially microscopic marsupial egg still gets encased in shell membranes and grows through absorbing uterine secretions. However, it then hatches *in utero*, implants in the uterine wall and – despite what mammalian nomenclature suggests – briefly forms a placenta.

Most marsupial embryos use only the yolk-sac membrane to form a placenta. Interestingly, though, not all marsupials have only yolk-sac placentas. In bandicoots, a placenta is formed using the allantois – the membrane that amniotes evolved to allow them to breathe through the surfaces of their shells (and excrete waste). And this is the same membrane from which eutherians form a placenta.

Marsupial placentas can also, like eutherian ones, influence maternal physiology, although some marsupial

---

[*] Consistent with decreased pre-fertilisation energy provision, most birds and reptiles have three yolk-making genes, but only one such gene remains in platypuses. It's also unlikely that paternal genes influence maternal provisioning in platypuses, and genetic imprinting has not been found in them.

mothers barely know they're pregnant – their physiology is hardly altered by the embryo's presence. One hypothesis for why marsupial placentas persist only briefly is that these animals have never evolved a mechanism by which the embryo is shielded from the maternal immune system. The accommodation of the alien embryo probably pivots on the invention of the decidual stromal cells in the eutherian lineage, although many things about the immunological tolerance of pregnancy remain uncertain. Marsupials, therefore, birth their tiny young and go about their milk-heavy reproductive ways.

In contrast, eutherians experimented with increasingly long gestation times. At some point, they ceased altogether to make a shell membrane, and the placenta grew ever more sophisticated, allowing prolonged pregnancies. While genetic conflict may fundamentally shape the nature of this pregnancy, it remains the case that the mother and foetus ultimately share the goal of a healthy newborn being born, and this dynamic reproductive strategy has self-evidently been a highly profitable system for placental mammals.

Much of people's everyday thinking about parenthood needs to be re-evaluated in light of Trivers' and Haig's conceptualisation of the process as fundamentally shaped by conflict between parents and young. But Haig has said there's been a reluctance to accept the idea. While no one questions that clashes of wants occur in adolescence, say – where they play out before us in moral and social spheres – the idea that conflict occurs from the moment an embryo implants in the womb is surprising and unsettling. It contradicts the dominant narrative of parents always putting the needs of their children before their own. But then haven't we always known that families can harbour a multitude of nuanced tensions? Weren't John and William Hunter's many successes and their ultimate estrangement both born of a mixture of fraternal support and sibling rivalry?

Early on, Haig coined an enduring metaphor by comparing the conflict between foetal and maternal genomes to a tug-of-war. A healthy pregnancy is one in which the rope stays taut – neither side can win, each must cancel the other out. And, of course, they normally do. This interaction has been sculpted by millions of years of adaptations and counter-adaptations; if either mother or offspring had a fundamental advantage, it would harm the species. And while conflict is rife, one does have to recall that the basic shared goal of producing healthy offspring unites mother and offspring.

Nevertheless, Haig has stressed that obstetricians should understand this evolutionary perspective and its implications. While natural selection normally yields beautifully adapted processes, human pregnancy is difficult and dangerous. Blame for this has often been attributed to our unusual upright posture. But Haig says the intensely intimate co-existence of two genetically distinct organisms blurs the usual medical conception of healthy versus pathological physiology. What is healthy for the foetus can be pathological for the mother, and vice versa. As examples, pre-eclampsia – a common and dangerous rise in maternal blood pressure towards the end of pregnancy – and gestational diabetes may each stem from a mother's susceptibility to the demands the foetus puts on her body. Foetuses release signals into the maternal bloodstream that increase blood pressure (to favour blood flow through the placenta), and they secrete signals that manipulate maternal blood glucose, such that the mother becomes insensitive to insulin. Novel approaches to treating these disorders have been inspired by the notion of genetic conflict.

Looking back at Mariana's birth, I'm forced to re-evaluate what I thought of that placenta. Would it, now, be ridiculous to admire my children for having made such organs? It probably would. But we *do* praise our bodies – we celebrate our immune systems when they battle off an infection, and our bones when a fracture successfully heals … What's certainly true is that as parents we quickly learn that our children have their own agendas and that they can rub up against our own.

As for Isabella's early birth, Cristina's obstetrician said she was unsurprised. 'We'd known something was up …' But the pathology lab found nothing wrong with the placenta. '*Why* was she early?' the doctor continued. 'We simply don't know.'

Birth

Soon after Cristina was admitted to hospital for Isabella's birth, a doctor arrived bearing a syringe. It contained a synthetic hormone that mimicked a natural one that normally spikes in late pregnancy to prepare the infant for birth. The doctor swabbed Cristina's thigh and injected it. The hormone's primary action is to prepare the lungs for the switch from sitting dormant and uninflated in a fluid-filled womb to breathing air.

When a mammal passes down the birth canal, it does in minutes what it took vertebrates tens of millions of years to achieve: it transitions from being an aquatic organism to being a terrestrial one. It is not only the lungs that must mature; an entire suite of organs and physiological systems, evolved for life on land, must awaken. As the lungs start to pump, the circulatory system must change. The kidneys and liver are suddenly in sole control of maintaining salt balance and detoxification. The digestive tract must now feed the baby. Immune function must surge. And, now, the baby must also maintain its own body temperature. Life without a placenta is tough.

CHAPTER SEVEN

The Milky Way

I should, before I properly start this chapter, make it clear that the following is not about breasts. Across nearly 5,500 species of mammals, only humans permanently house their mammary glands – actively functioning or not – in an enlarged chest. Scientists abhor an n of 1. When a trait of dubious utility is unique to a single species – which doesn't happen that often – there's nowhere to look for shared facets of biology that might explain how and why evolution created it.[*] If lionesses and female walruses also had breasts, then we might speculate that males with odd facial hair evolve a penchant for shapely chests. But they don't. No other species shares our possession of – and fascination with – breasts.

It's not that evolutionary biologists have ignored the challenge of the human breast. It's just a difficult question. Larger breasts don't produce more milk; breast size – being largely determined by the amount of fatty and fibrous tissue rather than the size of the milk-making gland itself – does not correlate with the capacity to lactate (although big ones may be fractionally better at storing milk). Breasts, therefore, are almost certainly signalling devices of some sort, and only tangentially related to the subject of this chapter – the evolution of a nutritious white liquid secreted from a mother's skin to feed her offspring.

Certainly, to see your partner become a mother is to witness whatever enigmatic beacons of sexual signalling might once have existed undergo a transformation on the scale of the industrial revolution. Twice now, I've watched Cristina's chest metamorphose to house two dedicated dairy

---

[*] Albeit the same is true of peculiar traits unique to larger clades; lactation itself, of course, being found only in mammals.

factories. And twice, I've experienced milk and its provision become an obsession.

My limited role has been to help deliver the fuel these milk-plants consumed. It's kind of suited me. In stressful times, I have a habit of cooking. Something, if possible, that requires stock, as if the aroma of bones being simmered might infuse those who need it with some degree of strength. When Cristina came home from the hospital after having given birth to Isabella, she arrived to a vat of minestrone soup.

Frequently while cooking, I contemplated the oddity of what we were doing to maintain our circle of nourishment: Cristina will eat this food, I'd think, process it in some way – which must surely expend a fair amount of energy – then expel a fascinating nutritive solution. Sometimes, I'd think of a pair of birds, the two of them, mum *and* dad, popping worms into their hatchlings' gaping mouths. I didn't feel useful enough.

But that was how it was. The girls needed milk. For their first six months, they consumed nothing but. Milk's ubiquity and our uniquely human way of never ceasing to consume the milk of other mammal species – on our breakfast cereal, in our coffee, cheese and desserts – diverts attention from the fact that for every other mammal it is a singular substance, a solution that evolved only to power life's first stages.

I suppose I've always known in some sense that in identifying myself as a mammal, I acknowledged my membership of a group of animals named for – and distinguished by – the females having mammary glands. But, except for one under-graduate essay on the 'ejection reflex', my thinking on this matter had gone little further. Watching a baby flourish on a diet of just one substance, I decided I would educate myself on how humans and our mammalian brethren came to rely on this arrangement. My conclusion has a familiar ring to it, but not for the usual reasons: mammary glands are amazing.

How do you like your milk?

Humans' unique breast anatomy is only one of many adaptations the mammalian lactational system has undergone.

Mammals may all feed their newborns milk, but the way they do so varies hugely.

Given their lengthy reliance on breastfeeding, it is unsurprising that the doyens of lactation are marsupials. The long relationships that develop between newborns and their chosen teats are underpinned by a unique physiology. In placental mammals, suckling induces the release of a hormone, prolactin, from the brain, which stimulates milk production in all mammary glands. But in marsupials, prolactin is constantly in the blood and suckling on a teat makes the mammary gland of that teat alone switch on expression of the prolactin receptor. Therefore, only suckled glands become responsive to circulating prolactin, limiting milk production to glands that harbour young.

After a newborn red kangaroo has completed her initial non-stop two-month suckle, she remains in the pouch, suckling intermittently, for another four months. Then, she splits her time between learning about the outside world and retreating to the pouch for some milk. Or sometimes just standing with her head in the pouch for comfort. When she suckles, she always does so from her chosen nipple. That nipple has grown up with her.

Integral to the prolonged lactational period of marsupials is a progressive change in milk composition. The milk's contents are suited to and, in part, direct the development of the young. A newborn kangaroo hauling itself up to its mother's pouch will sometimes find an older sibling already using one of the teats, and marsupial mammary glands being under independent control enables the mum to give each of these siblings milk tailored to their age.

Scientists have confirmed the influence of milk from different stages by cross-fostering wallabies. Young wallabies transferred onto teats that had previously been nourishing older wallabies started to develop faster on the new nipple.

Among placental animals, seals have evolved some impressive lactational strategies. For starters, the shortest known nursing period is that of the hooded seal. These animals live in the North Atlantic and Arctic oceans around

Iceland, Greenland and Canada, and they breed on floating pack ice. The females nurse their infants for a mere four days.

If you fear that this represents some negligent mothering, you'll be relieved to learn that over those four days the young seals double in size from their 22kg (nearly 50lb) birthweight, gaining around 7kg (more than 15lb) a day. And all that mass is at a loss to the mother, who fasts throughout the nursing period. This astonishing growth, coupled with the seals' cold-adapted lifestyles being based largely on being blubbery, makes it no surprise that hooded seals also have the fattiest known milk, around 60 per cent lipids, and it has the consistency of mayonnaise.

The advantages of hooded seals' wham-bam lactation system are thought to stem from the perilous nature of rearing young on floating pack ice. Unlike whales, dolphins and hippos, young seals cannot suckle underwater, and this necessary out-of-the-water platform is far from a stable childhood environment.

In contrast, fur seals – which belong to a separate branch of the seal and walrus family – mate and nurse on solid coastal ground and lactate for months. However, similarly to their hooded cousins, young fur seals are still fed very fatty milk in intense bouts of a day or two. The mother intersperses feeding sessions with trips to sea to forage and replenish her energy. In some species, these expeditions last for weeks, which should present a problem. In most mammals, when suckling ceases, the resultant build-up of milk in the mammary ducts signals that it is time for the mammary gland to return to its 'virginal' state.

Research in Cape fur seals has shown that their milk is missing a certain protein that is otherwise ubiquitous in milk. This protein can – in a petri dish, at least – induce the death of mammary-gland cells. To live their unique lives, these seals seem to have dropped a signal that normally leaches from accumulating milk to initiate the collapse of the mammary gland at weaning.[*]

---

[*] Or at least one such signal; research in rodents suggests there may be others.

Beyond seals, a mother bear will go up to two months without leaving the overwinter cave in which she nurses her newborns. While there, she eats not porridge but her offspring's excreta to stay alive. Lionesses in a pride, closely related as they are, will indiscriminately nurse any lion cub that is around. And wild pigs also cross-suckle their young. A sounder (one of those fantastic collective nouns) of wild sows even appears to synchronise pregnancies to favour such an arrangement.

Finally, whales. Whales and dolphins don't have lips and so have developed particularly muscular mammary glands to ensure milk is thrust securely into hungry mouths. And as usual, the magnitude of this operation in blue whales is off any readily comprehensible scale. From their two abdominal nipples concealed in banks of blubber, blue whale cows dispense milk almost as fatty and calorific as seal milk. And they do so at a rate of 220kg (485lb) a day. Over six months of lactating, the calf receives enough energy to feed 400 adult humans for the same period.

With all these mammals having such stereotyped nursing behaviours, one would think we'd be able to state categorically what is 'natural' for humans. Holly McClellan and colleagues, from the University of Western Australia, tried to do this in a study that compared lactation in 'two mammals that have evolved by natural selection, […] two mammals that have been domesticated and selected for special traits (the sow and the cow) and one mammal that is difficult to classify (the woman)'.

However, they concluded it was impossible. Across different human societies, the frequency at which women nursed, how much babies consumed and how long lactation lasted was all over the place. Even looking at traditional human societies – apparently free of the cultural disruptions that have warped the developed world – they found immense variability. The pastoral Khoikhoi peoples of south-western Africa breastfeed for only a few months, while Aboriginal infants sometimes suckle into their sixth year. Culture, it seems, gets everywhere.

On the origin of mammaries

Despite the diversity in milk composition and the duration of nursing, mammary glands themselves are remarkably similar across all living mammals. With even those of the nippleless monotremes having the same basic design, this homogeneity tells us that a sophisticated lactational system was in place prior to mammals as we know them existing. Consistent with this, the genes for making monotreme, marsupial and eutherian milk show that all the major components of this elemental foodstuff were in place before these three lineages diverged.

However, this also means biologists have no transitional form to interrogate – no animal has a half-formed secretory device that leaks something decidedly sub-milk. And like every other fleshy mammalian feature, mammary glands have left no fossilised impressions, and there are no cases of preserved milk, no cartons of Jurassic semi-skimmed frozen in the Arctic permafrost. So again we look to comparative developmental biology and genetics, and to inferences about the challenges mammalian ancestors once faced.

The first potential history of the mammary gland was published in 1872 by Charles Darwin. However, it was not Darwin himself who introduced the subject into evolutionary discourse. His thoughts on lactation appeared only in the sixth edition of *On the Origin of Species* as part of a long and careful response to what he considered the most extensive and damaging attack on his work.

That critique had come in 1871 in the form of a book entitled *The Genesis of Species* written by St George Jackson Mivart, an ardent Catholic whose PhD was awarded to him by the pope. Mivart was not an actual saint – he'd simply been named in full after the famous dragon slayer – and his concerns were not of a religious nature. A brilliant and erudite biologist, who specialised in primate skeletons, Mivart believed in evolution, albeit with the exception of it having shaped the human intellect. What he could not accept was the proposed *mechanism* of change.[*]

Mivart's chief objections were laid out in a chapter entitled 'The Incompetency Of "Natural Selection" To Account For The Incipient Stages Of Useful Structures'. And the mammary gland was high on Mivart's list of Useful Structures. He could not imagine a series of intermediate stages leading from no mammary gland to a suckled and prolific ductal system dispensing a nourishing liquid diet. Among his other examples were the wing and the eye. He focused in particular on how the earliest stages of evolving such things could not possibly have been helpful; how, he asked, could anything ever fly with 10 per cent of a wing? For lactation, he wrote, 'Is it conceivable that the young of any animal was ever saved from destruction by accidentally sucking a drop of scarcely nutritious fluid from an accidentally hypertrophied cutaneous gland of its mother?'[†]

Darwin took Mivart very seriously. The 1872 edition of *Origin* dealt point by point with issues he'd raised, Darwin offering what he saw as plausible scenarios for the stepwise evolution of various traits. For lactation, Darwin wrote that 'cutaneous glands, which are the homologues of the mammary glands, would have been improved or rendered more effective', so that certain glands 'should have become

---

[*] Mivart and Darwin had at first respected each other hugely and corresponded about evolution, but in 1873 they fell out bitterly when Mivart attacked an article written by Darwin's son George. More dramatically, Mivart was eventually excommunicated after he wrote a series of increasingly controversial articles about the Catholic Church. His books were placed on the *Index of Prohibited Books*, and in 1900 he was refused consecrated ground for his burial. His allies did, however, fight to have this posthumously overturned, successfully arguing that the diabetes that killed Mivart had for years been debilitating his mind.

[†] Mivart followed his concerns about the mammary gland with a one-sentence paragraph querying the usefulness of externalised male gonads. Alas, Darwin never responded to this challenge.

more highly developed than the remainder; and they would then have formed a breast, but at first without a nipple as we see in the [platypus]'. This pretty much described the next 150 years of investigating *how* mammary glands evolved.

But Darwin's speculations as to *why* the mammary gland evolved didn't have such longevity. At the time, most naturalists believed placental mammals had evolved from marsupials, and Darwin wondered if mammary glands had originated inside pouches. He likened such a scenario to an analogous feeding system used by seahorses for their pouch-reared young. He suggested that initially, a pretty indiscriminate skin gland leaked out some sort of fluid that the youngsters lapped up, and that those mammals with secretions 'in some degree or manner the most nutritious' would have prospered over their relatives, by having 'a larger number of well-nourished offspring'.

Today, we view marsupials and eutherians as cousins rather than successive stages in the perfection of mammals. And the pouch is thought to have arisen in marsupials after they split from placental mammals, suggesting that lactation evolved in pouchless ancestors. But more problematic was Darwin's focus on protomilk's nutritional value. In suggesting that milk's antecedent was a secreted solution that nourished the young, Darwin proposed that milk's precursor was a crudely leaked, less nutritious version of milk as we know it. Now, though, most authorities on lactation believe that protomilk or *proto*protomilk had a different, *non*-nutritive function.

Ironically, if this is correct, milk's evolution fulfils an alternative argument that Darwin used to rebuff Mivart. Acknowledging that evolving something as intricate as an eye, a wing or lactation was indeed an enormous task, Darwin argued elsewhere that perhaps earlier forms of complex traits wouldn't have had the same function as they do today.

Darwin illustrated the idea with an example from barnacles.* He contrasted two species: one that breathed by

---

* Darwin's book *Cirripedia* [*Barnacles*] resulted from eight years of gruelling first-hand anatomical investigation. In a letter written

gas exchange across the entire surface of its body and that had a pair of skin-folds to hold its eggs in place, and another that didn't grip its eggs (because of a secure shell) and that breathed through an elaborate lung-like structure. Darwin said 'no one will dispute' the similarity of the lung-like structure and the egg-holding device, and proposed that the lung had evolved from an egg-holder. That is, the barnacle equivalent of a lung didn't begin by incremental outgrowths where each stage sequentially improved the animal's ability to breathe. Rather, an outgrowth evolved originally to secure eggs in place, and only later did the surface area of that structure expand and elaborate (as the egg-securing role was dropped) to be utilised as a device through which to breathe.

As evolutionary biologists took up the challenge of explaining where milk came from, no subsequent major proposal for the mammary gland's origins suggested that nutrition was its founding purpose; if lactation was like barnacle-breathing, the search was on for the egg-holder equivalent.

And that might actually be a slightly better metaphor than it sounds. Since the confirmation that monotremes both lactated and laid eggs – meaning lactation's evolution predated that of live birth – and indications that the first mammals probably hadn't had pouches, theories of protolactation have focused on how an early abdominal secretion might

---

before the publication of the third volume, he said, 'I hate a Barnacle as no man ever did before.' Today, it seems a curious choice of topic, but in Victorian Britain there was apparently much fascination with marine animals, and Darwin's exhaustive four-volume study cemented his reputation as a naturalist. Plus, much as he hated the work, it directly taught him about variation, the bedrock of his great theory. In another letter, he wrote, 'I have been struck … with the variability of every part in some slight degree of every species: when the same organ is rigorously compared in many individuals I always find some slight variability.'

have improved the health of mammalian ancestors' eggs.* In fact, the egg–holder metaphor would be brilliant, if anyone still took seriously William King Gregory's 1910 speculation that a sticky abdominally secreted solution had stuck the eggs to the mother's undercarriage.

This idea followed the work of Ernst Bresslau, who'd earlier hypothesised that lactation had emerged after a highly vascularised incubation patch – an abdominal area kept warm by a dense weave of blood vessels – had evolved to keep eggs warm and had then started to release a substance that evolved to further help maintain mammalian eggs' elevated temperature. Problematically, only a seriously copious flow of heated solution would have warmed eggs rather than cooled them due to evaporation. Nevertheless, Bresslau and Gregory's theories remind us that some form of parental egg care must have evolved before lactation.

In the 1960s and 1970s, a number of further theories arose. First, J. B. S. Haldane, a giant of twentieth-century biology, suggested that mammals in a hot and arid environment *cooled* their eggs with water. Taking inspiration from Indian birds that provided their young with water dripped from their feathers, Haldane suggested that protomammals had used water from bathing to cool their eggs, and their hatchlings had then sequentially sucked this water from their mother's fur, then their sweat, and finally this sweat evolved into a nutritious solution.

Next came Charles Long and James Hopson. Long suggested that incubated eggs may have absorbed moisture or even nutrients from a secreted solution – recall that the monotreme egg is capable of accumulating secreted nutrients *in utero* – and Hopson discussed how becoming warm-blooded would have increased the likelihood of small eggs

---

* The main exception was the idea that milk began as a secreted pheromone that bonded young to their mothers. Lots of mammals have glands that secrete airborne chemical messengers, but they do so in tiny potent amounts that don't really seem like the basis of a meal.

and helpless hatchlings drying out, meaning they needed extra sources of fluid.

Elements of all these ideas live on, but today there are essentially two theories in play. Both of them are concerned with the health of protomammals' eggs, and in some ways they are not diametrically opposed to one another. The first was presented in the 1980s by Daniel Blackburn, of Trinity College, Connecticut, and Virginia Hayssen from Smith College, Massachusetts, and it has fared well since the molecular components of milk have become better understood. It suggests that milk began life as an antimicrobial solution.

The second hypothesis arose from the extensive work of Olav Oftedal, a scientist who is unavoidable if you look up mammary gland biology. Oftedal's synthesis of many aspects of mammalian and pre-mammalian biology holds that protomilk – along the lines of Hopson – was first important for preventing eggs from drying out.

I say the theories are at least partly complementary because it seems to me a minor leap to imagine that it would be a good thing if a solution that was moistening eggs was also killing any bugs attacking those eggs.

Olav Oftedal was born in Norway but has spent over 40 years in the US. His research – now done at the Smithsonian Institution in Washington, DC – has relentlessly pursued the biology of the mammary gland. He is responsible for characterising the lactational habits of numerous species, and he reports from first-hand experience that seal milk tastes 'fishy'. In seeking to explain the mammary gland's history, Oftedal has carefully woven together two main threads: the first is a detailed account of the anatomy of the mammary gland, and the second is an extended look at the challenges probably faced by our ancestors when they began doing something like lactating.

We'll start with the anatomy, and be warned: the following should be read without full reflection on the relationship between the subject matter and what you pour into your coffee, for mammary glands must necessarily be compared to some substantially less appetising glands.

A gland is any biological structure that secretes something. You have internal ones, such as your adrenal glands, and then you have surface, or cutaneous, ones. Mammalian bodies are covered in glands. Humans have over 40 different types. Appreciate, for example, that in your mouth, nose and ears you are currently making saliva, snot and wax. If it is hot you might be sweating. A combination of glands is keeping your eyes moist. Tiny glands at the base of your hairs are nourishing and waterproofing that hair. And if you are sexually aroused … well, no, you wouldn't be reading this book.

Of the cutaneous glands, mammaries, when functional, are the largest and most complex. Depending on the species, between one and more than thirty ducts run from openings in the nipple back into the breast tissue. These ducts are called, rather wonderfully, galactophores, and in the breast they split into branches that end in hollow sacs called alveoli. The cells that actually make milk line these alveoli, and, by various means, they secrete milk components into the cavities that they surround. Milk is expelled toward the nipples by layers of smooth muscle around the alveoli, which contract in response to a hormone – oxytocin – released from the mother's brain when an infant sucks on her nipples. This is the ejection reflex that I wrote about at university. I still remember being struck by how elegant the whole system was.

The question from an evolutionary perspective is which gland came first, and how they are all related to each other: which derived from which. The mammary gland has never been considered a candidate primordial gland modified to generate other types; so, the question is really which simpler gland transformed into the mammary?

Once again, the development of these structures has been central to unravelling from where they each came. The covering of the body begins as a uniform sheet of identical cells – the ectoderm – then local signals from cells just below the outer layer instruct little patches of cells in the ectoderm to switch identities. This happens all over the body, and it's not just glands that are formed this way – teeth and hair also arise from ectoderm.

After an initial patch of cells has formed, they start to assume the identity of one type of gland or another, and they morph and contort in characteristic ways. Successively, in response to consistently changing messages released by neighbouring cells, the constituent cells assume new identities, each more specialised and more resembling the adult form than the last.

For a mammary gland, the building process begins in the embryo along what are termed 'milk lines'. These are two mirror-image skin thickenings that run from either armpit to where human nipples lie and then straight down the belly. No matter how many nipples a mammal has and whether they're thoracic, like ours, or abdominal, as in cows, they all fall along these lines; mice have three pairs on their chests and two pairs on the abdomen.*

Mammary-gland development is a three-part story. First, in the embryo, along milk lines, patches of cells bulge out from the skin, then push back into the surrounding tissue, burrowing in and sending branches out through the underlying layer of fat cells. And then they freeze. They only develop further if they are in a female body, and then only when a cacophony of pubertal hormones instigates a second phase of expansion. A third and final stage is added only if the female falls pregnant; in response to more surging hormones, mammary glands then attain full functionality.

The comparison of milk glands with other glands has focused on the earliest stages of their development. Since Darwin, it's been believed that mammary glands are essentially hypertrophied sweat glands, but there's been a back and forth over which of two types of sweat gland they originated from: was it the eccrine gland, which throws out the watery sweat characteristic of humans but is otherwise rare among non-primate mammals? Or was it the apocrine

---

* On those infrequent humans with more than two nipples, their extras always fall somewhere on the milk line. The only exception to this rule I've encountered are those 13 opossum nipples which are arranged with 12 in a circle and one in the middle.

gland, which produces a more protein-rich sweat that comes out along hair shafts? Apocrine glands are rare in humans – they're most abundant in our armpits.[*]

Oftedal says there is no doubt about it: mammary glands come from apocrine glands.

An apocrine gland occurs as part of a triad. It is one-third of a complex with a hair, which it sweats along, and a sebaceous gland, which secretes sebum, an oily lubricant and water retardant for mammalian skin. When Ernst Bresslau – he of the egg-warming theory of lactation – looked at the early development of marsupial mammary glands, over 100 years ago, he found that the embryonic mammary gland was also part of a triad: it was associated with a hair follicle and a sebaceous gland. In kangaroos and opossums, the hairs were shed during early development to leave the nipple bald, but in koala bears, they were retained longer so that when the nipple first broke the skin, it had hairs still protruding from it.

In adult platypuses and echidnas, where milk seeps from their hairy lactation patches, the association between the mammary ducts and hair follicles remains robust; hair forms the tracks along which escaping milk runs. Given the observations in marsupials, this can confidently be viewed as the ancestral form.

Interestingly, apocrine glands, like mammary alveoli, are enveloped in a layer of smooth muscle cells, so perhaps, as mammary glands evolved, the mechanism for pushing out milk needed only to be tweaked rather than invented from scratch.

Developmental biologists are currently attempting to define the signalling pathways that direct mammary-gland

---

[*] In most mammals, this is the most abundant sweat gland, and some hoofed mammal species lack eccrine glands altogether. In addition to under our arms, we have them on certain parts of our anuses, genitals, eyelids, nostrils, ear canals and, indeed, around our nipples. Bacteria digesting the proteins that underarm apocrine glands secrete are the source of body odour.

development; part of this is pure research, part of it is concerned with what goes awry to cause breast cancer. There are many players, including the one that says 'Here will be a nipple' to induce the initial skin thickening, and the one that decrees 'Now cells, you will make a branching ductal system' to catalyse enfolding. It's really quite something to catch a glimpse of how nature builds itself.

As a former neuroscientist, it was funny to read that a molecule I knew as a regulator of synapse formation in the brain is also critical to mammary gland formation. Thrifty things, our bodies. The evolution of mammary glands is written in the evolutionary rewiring of these pathways. The main thing holding back evolutionary theory is a lack of equivalent knowledge about the development of apocrine glands ... Perhaps deodorant makers could be tapped for funding.

Germs and evaporation

With a feasible account of how sweat-dripping A morphed into milk-making B, we need to return to the question of *why* this upgrade happened. Let's stick with Oftedal for now and his idea that the forerunner of lactation was an increase in abdominal sweat production to prevent eggs from drying out. Oftedal thinks this happened well before genuine mammals existed, and that the spur for it came from our egg-laying ancestors becoming increasingly warm-blooded. The problem, Oftedal believes, was that these animals had leaky eggshells and that higher metabolic rates and higher operating temperatures would have put eggs in danger of losing catastrophic amounts of water.

Birds get around this problem by having hard, calcified shells, which are staunch barriers to water movement. But for some reason, the mammalian lineage appears never to have evolved such shells. Despite an abundance of ancient dinosaur and bird eggs, no one has ever found a fossilised egg laid by a mammal or mammal ancestor. Water-permeable eggs were probably the ancestral amniote condition, and they're not as much of a problem for cold-blooded animals – today, snakes, lizards and turtles faced with similar challenges can keep their eggs moist by burying them in soil or sand. But eggs at soil temperature would have been no good to mammalian ancestors increasingly geared to operating at elevated temperatures. Other modern snakes and lizards retain their eggs internally until they hatch. Again, this seems sensible, and as previously noted, this type of thing has evolved numerous times. But the ancestor of mammals didn't take this route. Perhaps certain protomammals did, but their fate was extinction.

Instead, Oftedal argues, the animals that were heading towards a mammalian destiny provided supplementary moisture to their eggs as they incubated them themselves. Certain living frogs and salamanders, while not warm-blooded, do this through mucus glands on their skin. And it would have been mammals' retention of glandular skin from their ancient amniote beginnings that was key. Through glands that first resembled apocrine glands, these animals would have released a fluid onto their eggs to ensure that the eggs did not dehydrate. From there, more and more specialised secretory apparatus would have evolved, and as tiny youngsters came to consume this secretion after they'd hatched, it gradually evolved into a food source.

Now, let's visit the antimicrobial camp. In two papers in 1985, Blackburn and Hayssen surveyed the components of milk to show that many of them were either similar to – or precisely the same as – compounds released from other skin glands for explicitly antimicrobial purposes. They proposed that the elemental step in evolving lactation was the selective advantage of an 'enhancement of egg survival through the antimicrobial properties of secretions of the incubation patch'.

Our bodies contain two types of immune systems to fight microbes. There's 'adaptive immunity', which is the clever one that learns to recognise and remember specific bugs so that responses to infections become ever more sophisticated as they proceed and you never catch the same cold twice. And there's 'innate immunity', an older defensive system composed of a number of mechanisms by which plants and animals attack or guard against foreign cells.

Some innate microbe-butchering relies on compounds contained in – *drum roll* – glandular secretions. Hayssen and Blackburn focused initially on the fact that alpha-lactalbumin – today a key player in the synthesis of the milk sugar lactose – is closely related to lysozyme, one of the key constituents of innate immunity secretions. The authors suggested that alpha-lactalbumin's molecular ancestor was originally in milk to help protect eggs from infections and then acquired sugar-making properties. They also note that other milk ingredients have antimicrobial properties. This theory got a substantial boost in the mid-2000s when further genetic studies found mammary glands and glands mediating inflammatory responses, in fact, shared numerous signalling pathways. Of particular interest was an enzyme known as XOR, which as part of an immune secretion slays bacteria, but when present in milk catalyses the formation of 'milk fat globules' – the means by which lipids are delivered to the young.

Overall, it seems that as protomammals became increasingly warm-blooded, their eggs could not be left unattended. An area of abdomen became highly vascularised for incubating eggs, and some glands there secreted a limited amount of fluid onto those eggs. This solution probably had advantageous antimicrobial properties, and increasingly warm (and small) eggs benefited from a more abundant secretion to stop them from drying out. Then the hatchlings started lapping up this secretion so that it was both egg-moistener and post-hatching food and drink. The secretion became more nutritious over time, constituents that once served only to kill microbes mutating to achieve this transformation. As the lineage moving towards true mammals contained smaller and smaller animals and offspring became more helpless, lactation was ever more vital. Finally, when live birth evolved, and eggs ceased to be an issue, milk was retained solely as a food – and, later still, nipples evolved.

Hayssen and Blackburn wrote in 1989 that 'the paucity of facts on the origins of lactation and milk renders any evolutionary explanation speculative', adding, 'the present ones included'. Developments in molecular genetics and developmental biology are helping to clarify things, but the origins of lactation will always remain buried in ancient history, a one-off experiment in maternal provisioning.

The food of love (and other things)

To draw a line here would not do lactation justice. A summary of why it was useful to evolve milk does not explain why milk was so useful a thing to evolve.

The most widely cited side effect of lactation on mammalian biology is probably its impact on our dentition. Most vertebrates generate new teeth throughout their lives, replacing worn-out ones with new ones as necessary. Sharks get through thousands in a lifetime. But mammals have only two sets: the so-called milk teeth and the adult set. The key feature of mammalian teeth – as we'll get to in Chapter 9 – is that, in adults, the upper and lower gnashers match up perfectly. This is possible because they grow in a mature-sized jaw. While the youngster suckles, its poorly occluding milk teeth are sufficient, then once its jaw is fully grown it can get to work on making its uniquely mammalian dentition.

But there's plenty more to think about than this. In 'The significance of lactation in the evolution of mammals', published in 1977, Caroline Pond, then of Oxford University, explored at length how this unique mode of feeding has shaped mammalian biology. And she concluded that lactation had been fundamental to a number of mammals' most profound adaptations.

Suckling mammals, Pond began, are supplied with 'a highly nutritious food source which does not need to be foraged for, masticated, digested or detoxified', all they need do is 'respire and excrete'. So, despite being physically separated from its mother, a young mammal can remain in a foetal-like state of nourishment and growth. And this arrangement seems to be mutually beneficial. If a mother is to continue providing for her offspring, it is much easier for her to track down food and dodge predators unburdened by the weight of these offspring. And for the young, it means their early postnatal growth isn't powered by their limited hunting or foraging abilities but by their mother's. This situation is quite unlike the daunting reality faced by, say, a reptile when it emerges from its egg and has to fend for itself.

Consequently, though, in terms of energy expenditure, this period asks more of the mammalian mother than pregnancy does. When pregnant, mammals only slightly increase their normal energy consumption – by about 10 per cent in humans – whereas while lactating, energy outflow increases by between 50 per cent (in humans) and 150 per cent (in rats). In fact, in many mammals, much of the extra energy consumed during pregnancy is used to lay down fat stores that are later converted into milk. Young mammals grow faster while suckling than they do *in utero*. And many mammalian young are nursed until they reach a considerable fraction of their adult size.

This arrangement means that mammals grow to reproductive maturity much quicker than equivalently sized reptiles. In a later article, Pond suggested that mammals are rapidly reproducing creatures good at colonising places, rather like the weeds of the animal world. This isn't quite the image of mammals I was looking for, but she has a point.

Lactation's centrality in the evolution of this pattern of rapid infant growth can, in fact, be seen in the fossil record. *Morganucodon* – a very early type of mammal that looked a little like a small stoat – left behind so many fossils 205–200 million years ago it has allowed for a fascinating analysis of its growth. Imagine if you measured how tall everyone in London was. The majority of data points would cluster around the average height of an adult human, but there'd be a splay of data from the under-18 populace indicating how these younger folks were still growing. Looking at all the data, you could infer that humans grow and then reach a fixed adult size – if they kept growing you'd see a very different-shaped graph. You could also draw some sort of conclusion about how quickly Londoners grow. If humans reached adult size by age 10, there would be a lot fewer sub-adult size measurements. Because *Morganucodons* were so common, enough fossil fragments have been recovered to do exactly this type of analysis, and the results showed that this early mammal grew quickly and to a fixed adult size.

But the fossils don't stop giving there. Comparing abrasion patterns on adjacent teeth, even fossilised teeth, one can see if those teeth are equally worn and, hence, the same age – or differently scuffed, indicating that individual teeth are newer than others. *Morganucodon* fossils reveal that these animals' teeth only got replaced once, suggesting that their rapid juvenile growth was milk-powered. By contrast, the size distribution of a slightly older ancestor indicated that it grew more slowly, and kept growing through adulthood. Its teeth? They were replaced individually and continuously.

Fast, milk-fuelled early growth, therefore, propels mammals through the perilous stages of early life and gets them to sexual maturity quicker – twin benefits favouring investment in lactation.

Another advantage of an infant's food arriving via its mother, pointed out by Pond, is that the mother can control when she consumes the necessary calories and when she releases them. Rather than a youngster itself – like a reptile – or a parent – like a bird – having to continually find fresh food to meet the dietary needs of the growing infant, a female mammal can lay down energy in fat and then dispense it when her infants need it.

This arrangement does two things. First, it protects the mammalian infant from random day-to-day fluctuations in food availability, keeping it a step or two removed from the dangers of starvation. Recent computer simulations suggest that such buffering would be especially valuable in times when food is only available intermittently and uncertainly. Second, it's a boon for longer-term reproductive strategies, permitting 'eat now, breed later' approaches. Picture that hooded seal losing 30kg (66lb) in the four days she nurses: that was an outpouring of resources accumulated over a much longer period. Or the mother bear who, because of last year's fat reserves, can shelter from winter in a cave with her newborns, so that the cubs learn to forage among spring's plenitude.

Throughout Pond's essay, she contrasts mammalian lives with mammary-gland-free existences, and she illustrates a final benefit of evolving lactation via Nile crocodiles. A freshly hatched Nile crocodile is only about 25cm (10in) long, but it has to feed itself immediately.[*] According to San Diego Zoo, adult Nile crocodiles eat 'almost anything that moves'. In the wild, they would typically eat antelopes, wildebeest and young hippos, dietary items unavailable to a 25cm-long anything. Instead, a crocodile hatchling first eats insects and occasionally frogs or snails. Then, as a juvenile, it progresses to eating rodents, birds and crabs. Consequently, crocodiles (and other reptiles – the infant and adult diets of lizards and iguanas being similarly distinct) must live in habitats that supply the multiple food types necessary for sustaining all their phases of life, which invariably restricts where they can live.

In contrast mammals, with lactation, could have just an adult diet and milk – the implications of which were pretty immense. It meant they could become very specialised feeders. Adapting to focus on just one food source made them more competitive in acquiring this foodstuff and opened up new habitats for them. They were free to explore all sorts of food-poor places that were simply not options for reptiles and their ilk.[†]

Pond even argued that lactation and the lifestyles it permitted might have been pivotal in the explosive rise of the mammals that followed the meteor that wiped out the dinosaurs. That meteor caused all sorts of climatic mayhem. And as food supply probably doesn't get much more intermittent and uncertain than in such conditions, milk may have been central to certain mammalian lineages surviving.

---

[*] Crocodile parents do guard their nests – being a Nile crocodile must help in this regard – but they don't feed their young. Parental provision is extremely rare among reptiles.

[†] Birds tend to be specialised feeders too, but some species that eat seeds as adults feed their hatchlings insects, which in part explains why many birds migrate to insect-abundant climes to breed.

In the next chapter we'll return to the fact that lactation demands that mother and infant mammals must form a placenta-free bond, and explore the additional exchanges that occur between generations of mammals.

Of mammaries and men

I've never forgotten the feeling I had when I first saw my partner Cristina breastfeeding our daughter Isabella, the sudden acute awareness of the limits of my biology.

Beyond human cases that clearly relate to something medically untoward, the only two serious claims of lactating males were made in the 1990s. One was a high-profile report on Malaysian Dayak fruit bats and the other concerned masked flying foxes (another type of bat) from Papua New Guinea. Microscopic analysis of male Dayak fruit bats showed that they had well-developed mammary tissue and milk ducts, which contained a small, but obvious, volume of milk.

Twenty years on, these reports are controversial. Concerns have been raised about the appropriateness of calling it evidence of lactation – a term that implies active food provision to youngsters. The male breasts contained just 1.5 per cent of the volume of milk a lactating female produces, and the male nipples were not 'keratinised' as suckled ones would be. Critically, males were never observed nursing. Maybe the fruit these bats ate contained high levels of compounds resembling female hormones. The jury remains out. And this being evolutionary biology, even if it were claimed that what was seen was the beginnings of male lactation, the jury would have to remain out for millions of years.

Whether male lactation has, indeed, had sufficient time to evolve in mammals was one of the questions Martin Daly of McMaster University, Ontario, asked in his thorough 1979 treatment of this topic. In 'Why don't male mammals lactate?', Daly first assessed if the bodily obstacles to male lactation were particularly insurmountable, concluding that the apparent physiological and hormonal tweaks required seemed relatively minor. Males, after all, have nipples

– unless they're rats, mice or horses – and rudimentary mammary glands form during embryonic development. There's nothing one can't imagine evolution coping with.*

Others have since argued that the necessary hormonal modifications might have off-target feminising effects that would make males less equipped to fulfil their masculine roles – something that could be a more serious issue for a gibbon defending his territory than for me stirring soup.

So what has stopped male lactation evolving? For starters, there's the fact that in 90 per cent of mammals, dads have nothing to do with their offspring. Then there's paternal uncertainty: with long internal gestations and frequent polygamy, there's a serious risk for male mammals that investing time and energy into breastfeeding would be heavily investing in someone else's offspring.

But then, many dads do parent. Paternal input has evolved independently across at least nine different orders of mammals. It's always in species that show monogamy – where males can be *relatively* confident of paternity – and that rear a small number of infants, perhaps only one at a time. From this, though, comes perhaps the major stumbling block for male lactation. In these cases where a fatherly helping hand *is* given – making breastfeeding dads a possibility – the provision of milk is rarely a limiting factor in the infant's growth, so the offspring would not benefit from a second set of mammaries.†

---

* Obviously, the 1990s fruit bats post-dated Daly, but he was strangely adamant that no reliable reports of lactating males existed. However, some rehabilitating Second World War prisoners spontaneously lactated due to hormonal imbalances caused by the structures that make and break down hormones recovering at different rates. Additionally, certain psychiatric drugs and brain tumours can also induce lactation. These reports, among other anecdotal ones, strengthen Daly's argument that it's not that difficult.

† This life strategy contrasts with a reproductive strategy based on the production of more offspring – litters of 10 to 20 in mammals – who get supervision and protection only from the mother, while

Finally, Daly confronted pigeons. When I'd once admired the egalitarian parenting style of birds – picturing alternating deposits of worms – I'd done so unaware that pigeons, both mum and dad, make a crude form of milk. Flamingos also do this, as do emperor penguins. This so-called 'crop milk' consists of cells sloughed from the birds' throats and apparently, if texture had determined its name, it might have been called 'crop cottage cheese'. It's not a widespread phenomenon in the avian world but in these few species it is robust, and is about as close as non-mammals come to lactating.[*]

Daly argued that in birds, co-parenting crucially predated the evolution of milk provision, so as crop milk evolved it did so in the context of a system in which fathers parented. By contrast, the original mammalian lactators were almost certainly single mums. Fatherly investment only evolved later and, in a minority of species, and it occurred in bodies specialised for being male. Therefore, fatherly contributions took other forms: territory defence, say, or foraging, or seeing off predators. On the flipside, the pre-existence of calcified shells probably meant that birds didn't need to protect their eggs against water loss when they became warm-blooded. Historical contingency is elemental to what can and does evolve.

The instinctive response to birds vomiting up crop milk is 'Urgh!' And it's easy to dismiss it as a weird trait of a few disparate feathered animals. But remember that mammalian milk appears to have evolved from a few animals sweating an antimicrobial fluid over their eggs. I can't help picturing a *Diplodocus* catching sight of our heroic little ancestor crouching to let her infants lap goo from her abdomen. Surely, he too would have thought, had he been able, 'Urgh, how very strange.'

---

males run around from one receptive female to another. Ironically, in such species, the necessary calories for milk production are often at a premium, but dad is nowhere to be seen.

[*] Although in November 2018, Chinese scientists reported that in a species of ant-mimicking jumping spider, mothers secreted a substance akin to milk from their egg-laying openings and their offspring fed on it for about 6 weeks.

On the subject of sex differences: isn't it strange, despite Western civilisation having been for centuries a straight-up patriarchy, that we are called 'mammals'? That this celebrated cadre of animals, clearly humans' closest cousins, is named after a female trait? And, for that matter, that this name is derived from a structure possessed by only half of mammals and that is functional – if ever – for only a fraction of its possessor's life?*

That Carl Linnaeus never offered an explanation for his choice has always left this open to speculation, especially given that lofty scientific principles didn't always guide Linnaeus's naming practices. A number of his teachers and mentors, for example, had attractive plants christened in their honour, while he gave the name of his fiercest rival to a weed. Londa Schiebinger, a science historian at Stanford, has looked most closely at the historical circumstances in which Linnaeus worked, and shown how he was involved in fierce sociopolitical battles over the practice of wet nursing. In Europe in the mid-to-late 1700s, wealthy Europeans were handing over their offspring to surrogate feeders at staggering rates, and Linnaeus – a practising physician and father of seven children – campaigned to halt the process. In 1752 – six years before he coined the term Mammalia – he published a treatise that attacked wet nursing on multiple fronts, believing that the practice was contributing to terrifyingly high levels of infant mortality. In this, Linnaeus appealed to the naturalness of breastfeeding, emphasising how even great beasts such as lions, tigers and whales offered gentle maternal care to their young and stating that Nature herself was 'a tender and provident mother'. For Schiebinger, Linnaeus's name choice stems directly from these political convictions.

As a scientist, I look at the niches opened by lactation in all its various forms, the avenues it opened for mammalian biology and, simply, the way it propels a new generation towards maturity, and think 'mammal' is a good name. And

---

* Germans, ever sensible, translated 'mammal' to *Säugetiere*, meaning suckling animal, something every mammal is for a period.

as a father, I look at my two young daughters and recall that everything they did for the first six months of their lives – every coo they cooed, every heart-dissolving, gum-laden grin they stretched – was milk-powered. I remember them at their mother's breast. I think how lactation was the midwife of so many unfathomable parental feelings. And I think, Carl Linnaeus, in my humble opinion, no matter what your motives were, if you'd called us the hairy ones, the four-chambered-hearted ones or the hollow-eared ones, you would have made a mistake.

CHAPTER EIGHT

Kids, Behave!

Rats are adept at pressing levers, especially lab rats that have been trained to know that pressing levers can earn them things. The frequency of lever-pressing is used by behavioural neuroscientists to infer what the animals value: the more presses, the more they want what the lever delivers. Rats will flick levers like crazy to get addictive drugs and sugary food, but if the reward is rat *pups*, there's a stark dichotomy in the way different rats respond. Males and virgin females couldn't care less for pup-giving levers, but new mums will rush to press them a hundred times an hour to fill their nest with up to twenty newborns.

To use the technical term, virgin females find pups aversive – usually, they simply avoid them, but occasionally they'll trample or attack them. When a rat becomes a mother, something profound happens in her brain to make her start embracing pups.

You may think, 'That's interesting. I can see how that could relate to humans.' But one has to be careful when extrapolating from rodents to humans – people are not simply scaled-up rats. Neurobiologists know of substantial differences in the way rodent and human mothers respond to youngsters (and, indeed, many virgin humans are very fond of babies). Nevertheless, when I encountered this research, I felt more like a rat than I ever have in my life.

I have a friend who, a decade on, *still* bursts out laughing when she recalls my attempts to hold – and to pretend to enjoy holding – her newborn son. I tried, I really did, but there just wasn't anything there. I was trying to imitate other people – those cooing 'Oooooh, he's so cute, I could eat him!' I was befuddled. You could *what*? You want to *eat* a baby?

Then, of course, I got my own daughter. I've said parenthood was transformative, but what was incredible was how *abruptly* transformative it was. The night of Isabella's birth, I entered the NICU to meet my hours-old daughter. When I pressed my sterilised finger into her tiny palm, and she wrapped her fingers around it, that was it. My perspective, my priorities flipped. No longer did I see the world from a single viewpoint. No longer was I the most important person in it. 'We thought we'd had quite a day,' I said to her, 'what a day you've had.'

Out of the womb, Isabella would interact with her parents differently now. In purely calorific terms, the mammary gland would be central, but a multitude of new channels for intergenerational exchange had just opened up for her.

Parental economics and the rare mammalian father

Since the 1960s, whether or not animals care for their offspring – and, if they do, how much care they provide – has been expressed in the hard, cold language of mathematics. Natural selection has no interest in how maternal or paternal devotion pulls at the heartstrings, nor should biologists. Parental care provides obvious advantages in yielding stronger, fitter offspring – animals that heighten the parent's chances of having grand-offspring. But the crucial issue for theoreticians trying to express nature's seemingly limitless variability in parental behaviour in mathematical terms is that such care comes at a cost to the parent.

Their equations lay out how the degree to which an animal parents results from a trade-off between costs and benefits. The benefits come in the form of the offspring's heightened survival and subsequent reproductive success, which are ultimately increases in the parents' overall reproductive success. The costs accrue from time and energy spent on parenting restricting the parent's chances to go off and breed further. From the interactions between these forces, natural selection finds a compromise between caring for offspring and trying to make more.[*]

---

[*] These arguments apply to every stage of parental investment, from allocating resources into egg yolks, building nests, caring for eggs,

It's from this type of work that Robert Trivers' ideas about parent–offspring conflict emerged, and his observations on weaning monkeys illustrate nicely how costs and benefits affect the mother. While the mother lactates, she is benefiting herself by aiding the suckling offspring's development, but this comes at a cost to her because while she provides milk, she does not produce another infant.

A key aspect of this is that the costs and benefits are not constant: they change over time. Providing milk to a newborn monkey (or any mammal) is essential – the benefits are absolute. But as the youngster develops and can increasingly fend for itself, the benefits of lactation diminish for the mother until her investment in milk-making outweighs the payback she gets from it.

And this doesn't just apply to monkeys. Across all mammals, theory suggests that how a species lactates – all the variety we encountered in the last chapter – will be shaped by a set of costs and benefits that change according to the particular biology of the animal in question. For example, mammals with access to abundant food in their breeding seasons produce, on average, more milk per day than those with scarcer resources, but the latter nurse for longer. The hooded seals that pump out 7kg (15lb) of milk a day for just four days are responding to the singular realities of living on pack ice in Arctic temperatures. The orangutan nursing her daughter for six to eight years does so in the context of a richly social group of primates, where the value of a mother–daughter relationship stretches beyond the purely nutritional.

And so it is with any aspect of parental investment: whether parenting occurs, and to what extent, will be determined by the exact circumstances of the species under consideration.

---

retaining embryos in wombs, through to lactation and aiding youngsters who are quite capable of feeding themselves. Here, though, we'll focus on postnatal (or post-hatching) care, taking it as a given that in mammals natural selection has already acted to produce a reproductive strategy founded on heavy prenatal investment in a limited number of offspring.

Casting a wider net, the sporadic evolution of parenting in fish, amphibians or reptiles, for example, occurs in species living in dangerous or unpredictable environments, frequently where predators abound. Under such circumstances, the benefits of parents guarding their eggs or young are large enough to outnumber the costs.

For mammals and birds, parental care of newborns and hatchlings is non-negotiable. The demands of maintaining a warm-blooded physiology mean that without parents providing nourishment, and typically warmth, these youngsters would die. In every species of mammal – and the vast majority of birds – the mother attends her young. In purely energetic terms, it's interesting to note that juvenile mammals, therefore, run at an energy budget deficit – they gather less energy than they consume – whereas mothers work to acquire more energy than they need to maintain themselves alone.

But then we come to the big difference between birds and mammals: in 90 per cent of bird species, the mother is helped by her male partner, while, as previously noted, in about 95 per cent of mammals, fathers do zero parenting. The high prevalence of paternal care in birds is usually attributed to the hatchlings' energy demands needing to be met by two adults. Often birds forage over considerable distances for widespread food, and the mother cannot store previously gathered energy to supply it as food as the lactating mammalian mum can. Plus, predators can easily access avian hatchlings, so having one parent guarding and one foraging has clear advantages.

Given how most of us human fathers like to consider ourselves useful, the situation in mammals is almost embarrassing. But again natural selection doesn't care what we think. The two questions raised by the mammalian dearth of paternal care are: first, why don't fathers, as a rule, participate? Then, second, how come 5 per cent do?

For starters, internal fertilisation in general acts against paternal input. When a male stickleback puts his sperm onto some eggs he's just seen being released, he knows that the young he is about to guard are his. A male mammal seeing a female run off after doing the deed with him has no idea what she'll do next, which rather limits his certainty that any offspring she subsequently has are his. And if the male isn't sure of his paternity, he won't invest care.

On top of this, the way in which reproduction has evolved in mammals has also made males increasingly peripheral figures. The help a male can offer a female while she's carrying foetuses or feeding them with milk is limited. The benefits of paternal input can, therefore, be rather small. Are two partners guarding a litter much better than one? Can a father teach the young skills the mother can't?

And in mammals, the benefits of paternal care need to be large, because, for the male, the *costs* of it are huge. To be blunt: say a woman slept with 10 different men in a month, and that a man had sex with 10 different women in the same timespan. Whereas the former would not increase the number of offspring she might have, the latter might father 10 babies – the male's reproductive potential increases with notches on the bedpost, but a female's doesn't.

This difference in the rates the two sexes can reproduce exists in nearly all animals, owing to the nature of making big eggs versus producing loads of cheap, quick sperm, but in mammals it's especially pronounced. The discrepancy profoundly influences the respective sexes' reproductive behaviours and, for male mammals, the potential rewards of heading off to find new mates means that staying put comes at a considerable cost.

Reflecting on the evolution of behaviour can be difficult for us *Homo sapiens* because we tend to consider behaviour volitional. We think of brains as organs that allow animals to *decide* how to behave. But the male platypus is not actively choosing to post-coitally abandon the female because there are more tails out there to be bitten; baboons don't deliberate over how they might breed and parent, and rats don't rationalise their reproductive strategies. The behaviour of non-human species can vary to some degree, but its central functions are fixed. In 95 per cent of mammals, natural selection has simply propagated males that sought further reproductive encounters, at the expense of those who stuck around to try and help out with the kids.

Why then is it different for human males and another 5 per cent of mammal dads? The first prerequisite for paternal care seems to be that the males are in a monogamous relationship with the mother. Paternal care is only known to occur in three species where the mother and father are not going steady – one mongoose, one marmoset and one lemur.

Mammals live in many social arrangements with many mating patterns. Sex ratios in groups vary; males might mate with multiple females; females can mate with one or more males; social rank may be the key; and males often engage in all sorts of antler-cracking, head-banging, teeth-baring, scrotum-flashing, stinky secretion-wafting, neck-bashing, howling or boxing confrontations to gain access to females. Sexual encounters may happen at very specific times in the season – female porcupines, for example, are apparently receptive for only 12 hours a year – or it may happen across the whole calendar. For all these many arrangements, theory posits that a pattern of costs and benefits to males and females accounts for how things get done. And somewhere among this cacophony of sexual strategies lies a small minority of monogamous mammals.

There have been two main suggestions to explain the evolution of monogamy in mammals. One is a matter primarily of geography, while the other concerns the ugliest of mammalian behaviours.

Males of many different mammal species will kill the offspring of other males to breed with the dead offspring's mother. Infanticide is alarmingly common. In some species, it's the principal cause of infant mortality. It happens primarily in species where the death of her offspring brings the female back into heat, allowing the killer to mate with her. Hence, one theory of monogamy posits that it evolved as a male strategy for defending against infanticide.

It's possible too that paternal care – other than defence against intraspecific infanticide – has positive enough effects on offspring survival or development to make monogamy with paternal care beneficial above the lost mating opportunities for the father.

The other view is that monogamy evolves as a mate-guarding strategy when females are especially spread out geographically or 'intolerant' of other breeding females. In this case, the male is ensuring that no other male gets to breed with *his* partner.

In a bid to distinguish between these theories, Dieter Lukas and Tim Clutton-Brock of Cambridge University surveyed an impressive 2,545 species of non-human mammals and plotted their mating behaviours and levels of paternal care on the mammalian family tree. This 2013 survey allowed Lukas and Clutton-Brock to deduce the conditions from which monogamy had evolved in 229 species, 9 per cent of the total.

The first thing to note is that in 94 of the 229 monogamous species, males contribute nothing towards childcare. For example, among dik-diks – diminutive antelopes that live in southern Africa – males are completely monogamous but do not guard, feed or teach their young.* Importantly, the fact that there were more monogamous species overall than there were monogamous species where males provided paternal care suggested that fatherly care only emerges after the evolution of monogamy. The alternative possibility had been that paternal care would have evolved first – outside of a monogamous relationship – then led to exclusive relationships forming between the mother and father, in which case you'd have expected to find more species that showed paternal care but were not monogamous.

That the data said couples formed first, and males started to parent second, made Lukas and Clutton-Brock look for the factors that favoured the evolution of monogamy. They found that pair-bonding developed in species where the females were spread wide and far, leading solitary lives in large ranges. When females are distributed this way, it seems that males are unable to defend multiple mates and that their best reproductive strategy is, therefore, to pair up with a single female. As one commentator put it, 'female mammals

---

* The name dik-dik apparently echoes a sound the antelopes make, rather than being a comment on this apparent neglect.

set the ground rules, and males map themselves onto their distribution.'

In only one species, a lemur, did it appear that monogamy had evolved from social living. Whether humans would represent a second such instance is an open question.

The pattern of paternal care among mammals shows that it is most common in primates and carnivores but also dotted about in other orders. Wolves and African hunting dogs are often diligent fathers, returning to their young with chewed-up meat for them to eat. Young dusky titi monkeys are carried and protected primarily by their dads, only going to mum for grooming and milk. Helping make the case for fathering, Lukas and Clutton-Brock also found that monogamous couples in which the father helped out in some way produced more litters each year than either monogamous couples with uninvolved fathers or solitary females.

Bonds

To return to how parenthood changes a mammal, let's get back to the brains of those new-mum rats and the transformations they undergo.

Rat pregnancy – like all mammalian pregnancies – is a marvel of adaptive physiology. At the heart of the process is a web of hormones. The main players are the sex steroids, progesterone and oestrogen, and two peptide hormones released from the base of the brain: prolactin – pivotal, as the name suggests, in milk production – and oxytocin.

The primary source of these hormones is the placenta, meaning that foetal genes again play their part in gaming the system in favour of the offspring. We'll proceed, however, under the safe assumption that both mother and young share the common goal of healthy offspring emerging. These blood-borne messengers first control the uterine environment, regulate maternal energy storage and consumption, and prepare the mammary glands for action. Then, when the time comes, they orchestrate the infant's exit. Not that reproductive hormones are then finished: after birth, they keep flowing to regulate lactation, maternal metabolism and the mother's brain.

In fact, during pregnancy hormones act on neural circuitry so that mothers behave in accordance with the demands of childrearing. In many species, for example, mothers will build nests for their imminent arrivals. And as pregnancy ends, hormones 'prime' the brain for full-blown motherhood, this priming making the brain responsive to the biochemical tsunami that induces labour and birth.

The hormone oxytocin gets its name from the Greek for 'sudden delivery'. A surge of it is the trigger that stimulates labour – its actions on uterine smooth muscle have been known for a long time. But, in 1979, Cort Pedersen and Arthur Prange, at the University of North Carolina, injected oxytocin into the brains of virgin rats. The result? Many of them started to act maternally. And this behavioural change happened more frequently if oxytocin followed a priming course of oestrogen that partially mimicked being pregnant.

We now understand that oxytocin and oestrogen, plus certain other biological messengers, act together in a brain region called the hypothalamus. There, they increase signals to the brain's 'reward circuits', the rejigging of which alters the response a rat pup evokes in the mother.

Reward circuits are rudimentary parts of a brain, functioning to give objects or animals in the outside world value – either negative or positive. It's these circuits that make an animal either avoid or embrace the things it encounters. Some aversions are innate: mice, for example, are instinctively wary of fox urine. But attached values can also change. In a maternal brain, the reproductive hormones seem to combine both to reduce the virginal aversion to pups and, in parallel, to activate brain circuits that make pups attractive. The result of the oxytocin surge is, therefore, a mother that immediately bonds with her newborn young.

Oxytocin has subsequently had a remarkably high profile for a molecule. Promoting the simple idea that oxytocin can melt away malevolence and ramp up an animal's affections, the hormone has been called the 'love molecule', the 'hug hormone' and the 'cuddle chemical'. One news-splash study reported that a quick squirt of oxytocin up the noses of people playing a game made them more trusting. But, as with all attempts to pin complex behaviours on the abundance of a single molecule, the story is not that straightforward. Oxytocin is now viewed more as a chemical whose release indicates that a social event of potentially great importance is happening, and it helps the brain respond more fully to the sights, sounds and smells that accompany such an occurrence.

Its actions on the maternal brain's reward circuits are profound, but it functions in concert with signals from the pups. For instance, oxytocin modulates activity in two brain regions concerned with the mother learning to recognise the scent of her litter. In rodents, the sense of smell is socially very important, and information about odours feeds into the reward circuits. Additionally, a 2015 study showed that oxytocin made maternal mice more responsive to their newborn's ultrasonic cries by acting on hearing centres.

So do these hormones account for my seeing similarities between rat mums and how I changed when I first held my daughter's hand? Alas, probably not. For one, I hadn't been subject to pregnancy's hormonal roller coaster before I met Isabella. And two, when it comes to the induction of paternal behaviour there's some very interesting variation across mammals.

You can't study paternal behaviour in male rats, as they don't participate in their offspring's upbringing. But researchers interested in paternal care and the experimental utility of rodents have taken advantage of fatherly care's presence in Mongolian gerbils, Djungarian hamsters and Californian mice. Studies of these animals – and to a lesser extent, paternal primates – have shown that male hormonal levels also shift with the advent of fatherhood: an increase in prolactin is commonly observed and, interestingly, testosterone drops. However, across species, male hormonal changes are much less consistent than female ones, and they have not been robustly linked to changes in behaviour. Perhaps, given that fathering, unlike mothering, is not a constant in mammals, its independent emergences in different lineages may have been achieved by different mechanisms. Also, male hormonal shifts might act to change a father's physiology or behaviour in subtler ways than the females', changes that are specific to the care they provide. Certainly, paternal contributions are diverse, from the up-close care of dusky titi monkeys, through male wolves bringing back food, to less direct input such as gibbon fathers defending their family's territory.

Studies of human males have confirmed that in us, too, testosterone levels plummet when we become fathers. And they remain low in dads who care for their kids for three or more hours a day, suggesting that the biology of human males really is designed for paternal care. And that brings us back to differences between rats and people. The detailed neuroendocrinological studies that have been done on rodents to figure out the basis of their maternal bonds have been done on just a few other mammals, including primates and, curiously, sheep. The picture that's emerged is that maternal bonds have been retained – and are fundamental – but their mechanistic underpinnings have meandered to suit the animals concerned.

In small mammals with small brains – which likely best reflect the original mammals – postnatal care can be viewed, in some regards, as an extension of pregnancy. The young are born deaf, blind and unable to move meaningfully, and the mother delivers them into a nest that functions to create a warm, moist microenvironment. Her main tasks are then to provide milk and warmth, and to retrieve any youngsters that crawl out of the familial home. The mother doesn't distinguish between members of her litter, and care typically only continues until weaning. There are meaningful behavioural interactions, as we'll find out, but they come only later.

By contrast, a wildebeest calf is on its feet in two or three minutes, and within five the calf is running with the herd. Prowling lions and cheetahs negate the luxury of a leisurely development. Shadowing its mother is essential for a calf's survival: most who become detached from their mothers wander, bleating and vulnerable, before ending up in the bellies of said feline predators. In wildebeest and other ungulates, mother–infant bonds must be formed instantly. For most larger mammals' newborns (humans are notable exceptions), early life does not involve lying around in a nest.

Though not quite as impressive as wildebeest calves, newborn lambs are also immediately able to maintain their body temperatures and to move about. Hence the interest in how sheep form maternal bonds. Barry Keverne and Keith Kendrick have shown that the reproductive hormones of pregnancy remain central to maternal behaviour. Virgin sheep, for instance, primed with oestrogen and progesterone, likewise become motherly when hit with a shot of oxytocin. Smell too is important in a ewe bonding with her lamb. But, crucially, within an hour or two, the ewe forms a very selective bond with her offspring and is aggressive towards other lambs. Neither sheep nor wildebeest mums will foster other youngsters. To recognise an individual rather than a youngster *per se* means that social odour processing in ungulates is more complex, discriminatory and requires more brainpower.

Reviewing the primate and human literature, Keverne finds even more differences with rodents. Babies certainly smell good to people, but the role played by this sensory system has diminished, while visual recognition has become more important. Most significantly, maternal behaviour in primates is no longer a slave to the hormones of pregnancy – for instance, caregiving often occurs in females who have never been pregnant. Keverne describes an 'evolutionary progression away from hormonal-centric determinants of maternal behaviour to emotional, reward-fulfilling activation' and talks of 'emancipation' from hormones. The brain's reward circuits remain critical, but, in primates they are engaged more with the cerebral cortex, which is enlarged in primates and huge in people. This diminished role for reproductive hormones may be an adaptation, at least in part, to primates providing maternal care for a long time past pregnancy, for months or even years.

If human mothers have been most emancipated from their reproductive hormones, using more of their brains' computing power in concert with the circuits controlling emotion, this seems to leave more room for males. Perhaps – helped by a precipitous fall in that blunderbuss hormone, testosterone – we men can develop attachments by similar means, attachments that still daily blow me away. It seems, also, to go some way towards accounting for the bonds my daughters have with their grandparents, and the exceptional care they've received from the nannies we have employed. I think too of my friends who adopted two infant girls. In natural selection sifting through costs and benefits, maternal care has remained fundamental in all mammals. But beyond this, we're left to contemplate a process that can yield both infanticide and the flourishing of my friends' daughters.

Early learning

Parenting, though, is no one-way street. Birth transforms the mother, but the newborn mammal is equipped, too, with instinctual behaviours that bond it to its caregiver. Typically, infants can quickly discriminate their mothers from other adults, and they attempt to stay in proximity to mother. Most young protest if left alone and show despair if such a situation persists. From rats to dolphins, mothers and infants commonly seek to maintain physical contact with each other.

Milk may be the cornerstone of postnatal maternal care – and, in some species, such as certain tree shrews, the mum barely has contact with her young besides feeding them – but, as Caroline Pond emphasised, lactation creates a union between two animals within which learning can occur. Studies in many mammalian species have shown that maternal care is essential for healthy psychological development.

Mammals – as we'll discuss in Chapter 12 – have large brains and the ability to learn a great many things, and consequently, mammalian behaviour isn't confined to inflexible instincts. Instinctive behaviours are seen in all animals. They are hard-wired responses to specific stimuli – as invariant as a mammary gland ejecting milk when it's suckled. Their adaptive value is usually plain to see, and crucially they don't have to be learnt. For example, mice raised in labs where they are unable to burrow will, as adults exposed to their natural environments, dig the burrows or tunnels typical of their species. But more advanced nervous systems enable animals to learn and stretch their behaviour beyond the rigidity of instincts.

There are a number of ways to learn. One is to give things a go and see what happens. Such trial-and-error learning was instrumental to the growth of behavioural psychology in the middle third of the twentieth century. Often using levers to allow lab animals to titillate their brains' reward circuits, research focused on the idea that animals learnt by monitoring whether their actions had positive or negative consequences: actions that brought rewards were retained, and those with unwanted effects were abandoned.

Trial-and-error learning is real and important, but it is a lonesome, time-consuming activity, and one that demands the animal learns everything from scratch. This can be dangerous. A baby howler monkey may have learnt that boa constrictors are to be avoided, but as he's being constricted this will be of little future use.

Identifying predators is a skill much better learnt from elders who've been around the block. Young monkeys and kangaroos, for example, do this by observing the alarmed reactions of adults to certain intruders. Such social learning allows behaviours to pass between animals and between generations.

Nowhere is such transmission greater than in our species, from cultural diversity, through Erasmus Darwin's influences on his children and grandchildren, to my daughter repeating swear words I accidentally say. It defines our species, but it's seen throughout Mammalia. Young rats, for instance, learn food preferences from their mothers. Although these animals will eat just about anything – part of the reason rats have thrived in humans' detritus-strewn wake – they're very conservative about trying new foods. Pups first get a taste for what their mum likes eating by flavours passing into their milk. Once older, the rats follow their mums on foraging missions, and the mother also marks palatable food with a chemical signal. This way, young rats end up following the dietary ways of their clan.

In Israel, received feeding instruction has even allowed rats to invade a new ecological niche. The floors of pine forests are barren places – the only nutritious thing on offer is the pine seeds. But these are buried inside cones. Squirrels know how to extract pine seeds; rats typically don't. However, a paucity of Israeli squirrels has enabled rats to take up residence in the pine forests. There, they strip the scales off pinecones and munch away on the seeds.

Opher Zohar and Joseph Terkel at Tel Aviv University showed that adult rats not living in these forests rarely learnt the complicated routine needed to strip a pinecone. And housing them with accomplished cone-strippers didn't help them. However, if a pup was fostered by a cone-stripper mum from the forest, it quickly became adept at accessing the seeds. This unique colony of forest rats, founded presumably by an adult that stumbled on cone-opening techniques, is sustained by an intergenerational transfer of behaviour.

Intuitively, you might imagine that the youngster imitates its mum – we humans tend to see copying as the heart of cultural inheritance. But imitation is actually a very cognitively demanding process. To mimic an action is to take a visual representation of someone doing something and to mentally transform that information into a set of muscle commands that recapitulate it. How common strict imitation is and which animals are capable of it are moot. However, social learning doesn't have to involve imitation; an animal can, instead, learn from observing others what features of its world are worth attention and also how the environment can profitably be manipulated. For instance, most of the non-forest rats paid no attention to the pine cones or just indiscriminately gnawed them. The pups, though, learnt first that cones were where food comes from and they were then directed to the right end of the cone to start trying to strip it. Only then did they probably employ trial-and-error learning to reap their reward. Other mammals similarly transmit learned behaviours within social groups: chimpanzees teach their youngsters to use sticks to 'fish' for termites, and Japanese macaque monkeys learn to wash sweet potatoes in the sea before eating them. In the 1980s, humpback whales were seen passing on to one another a new technique for rounding up fish.

Social interactions in mammals can reach levels of complexity seen in few other vertebrates – for many species sociality is a defining characteristic. However, Lukas, and Clutton-Brock's 2013 analysis of mammalian breeding systems indicated that the original mammals were likely solitary creatures. And most species still are, so we cannot say that group living is the mammalian way.

The numerous times social living has evolved – and it's found across various mammalian orders – groups most frequently form around related females, while males disperse to look for new mating opportunities. And sociality is most common in larger, and therefore larger-brained, species. It is most prevalent among cetaceans, carnivores and primates, although this rule is not absolute. Indeed, the species that come closest to insect-levels of sociality – that sort of 'colony as superorganism' type of organisation – are rodents. Naked and Damara mole-rats live in subterranean networks with non-reproducing colony members working in the service of queens. Naked mole-rat queens produce some of the largest litters of any mammal, typically birthing around 11 pups but sometimes as many as 28.

Attempting to understand when and how social groups evolve is again a pursuit written mainly in mathematical considerations of costs and benefits. Countless debates have arisen from the need to account for animals that apparently sacrifice their own well-being for the good of the group. Benefits, in general, can accrue from greater vigilance against predators, better food capture or defence, buffering against physical factors, increased breeding efficiency and, indeed, facilitated social learning. A lone lion cannot tackle a buffalo, whereas a pride can. The preyed-upon, living in groups, can split their time between scanning for predators and foraging, and better defend themselves: musk oxen, for example, form circles with adult males on the outside and the young inside. Small mammals often huddle together to ward off the cold. Costs associated with group living can stem from increases in disease or parasite spread; amplified competition for resources *within* species, with social groups sometimes aggressively clashing over food. Additionally, low-ranking members of groups can lose out on mating opportunities.

Primates have both the most complex social dynamics and the longest maternal care. Often, mothers play active roles in establishing their daughters' social ranks in the groups, which can be important for getting the opportunity to mate. Female vervet monkeys whose mothers remain in the group have more young than those whose mothers don't.

Monogamy too is a type of social bond (love and romance are not terms you encounter in zoological literature). And where the neurobiology of bonds between monogamous mammals has been studied, male–female attachment appears to operate on foundations built for establishing bonds between mothers and their offspring.

Play

This book would not exist if I didn't participate in a game where huffing and puffing humans try to move a ball into a net without using their hands. Isabella currently indulges me with brief football kickabouts, but she prefers crafts. Mariana likes mothering her dolls. They both dance, doctor, and together they run an excellent café; unless, of course, they're too busy winding each other up. Humans play. But far from being something only we do – passing it on to our dogs and cats, too – all mammals play. Like dogs; kangaroos, bears and rats play-fight. Pronghorns and other ungulates spar. Fruit bats chase and wrestle. Seals loll about in shallow waves. Juvenile ibex and mountain goats mess about on rock faces high enough to kill them if they fall. In the water, hippos turn backflips. Japanese macaques bash stones together. And grown bison run onto frozen lakes and slide across them apparently howling in delight.

At the broadest scale, the mammals that play the most tend to be the ones with the biggest brains, such as primates, elephants, cetaceans, ungulates and carnivores, although many rodents also mess about extensively. At the level of species, however, the relationship between play and brain size breaks down. Instead, in rodents and primates, the duration that animals remain juvenile better predicts play complexity.

At certain times, people have made a case for play only being present in mammals and certain birds. But this might have resulted from excessive caution about anthropomorphising animal behaviour. Since non-human play was put back on a sound scientific footing by Robert Fagen's 1981 book *Animal Play Behaviour*, it's been observed all over the place. Turtles, crocodiles, octopuses and wasps all play to some extent. So, again, we're back to a trait that is very highly developed in mammals but not exclusive to them.

One of the difficulties with play is defining what exactly it is. It seems to belong to that category of slippery phenomena that evoke Supreme Court Justice Stewart's method for recognising pornography: that is, we know it when we see it. This, however, won't do for serious academic studies. The University of Tennessee's Gordon Burghardt has devised a five-point scale for recognising play:

1 Play is incompletely functional – it doesn't really achieve anything obvious.
2 It's spontaneous, voluntary and rewarding. Play researchers, more than any other academic group, have to wrestle with what it means to 'have fun'.
3 It's clearly different from 'serious behaviour'.
4 It's repeated – but not in a pathological way.
5 Stress inhibits play.

What's also clear is that play comes in three forms: lonesome locomotor activities, object play and social play. But should researchers look to explain each form of play separately, or is it possible to unify all play – from doodling to play-fighting – as a singular process that serves a shared purpose?

In 1872 Herbert Spencer proposed that play was for warm-blooded animals to work off extra energy. Whereas in 1898 Karl Groos suggested that the playing animal is rehearsing and honing behaviours for the serious business of adult life. One of play's most defining features is that it is typically confined to specific immature phases of an animal's life. As investigators continue to debate why play exists, these two themes still resonate, although they may not be mutually exclusive. Play's occurrence day-to-day may be favoured by greater reserves of energy, while the activity's ultimate function may be to yield more successful adults.

Proving that playful young make more successful adults – that play has clear survival benefits – has been tough. Although, there is now supportive correlational data from wild populations of ground squirrels, bears and horses. Studying play in the lab is even harder. Researchers can prevent young rats from playing by isolating them from other rats, but this has a whole raft of confounds and just how many of those problems are alleviated by giving experimental subjects companions who've been drugged so as to be unable to play is debatable.

The problem with Groos's practice idea in its simplest form is that mammals prevented from playing as youngsters still display species-specific behaviours as adults. Kittens, for example, deprived of objects to manipulate, hunted as normal when mature. Other investigators have suggested that play peaks in periods when brain development is susceptible to experience, and that playing fine-tunes the wiring of that organ. Such theories have often focused on brain regions involved in movement control and finesse. Other ideas highlight the importance of play for moulding social interactions.

Focusing solely on mammals, Marc Bekoff, Marek Spinka and Ruth Newberry proposed in 2001 that the unifying principle of play is that it prepares the animal for the unexpected. They argued that mammals frequently aim to lose control of situations in play. By putting itself off-balance or out of position, or handicapping itself regarding sensory input, the playful mammal's goal is to respond to this loss of control. Real life's unpredictability lies at the heart of this theory: predators, prey and sexual rivals will come in a multitude of different forms behaving in unknowable ways, uneven terrains will be encountered, accidents will happen. Play – especially with self-handicapping – helps the young animal become a more dynamic adult.

Bekoff, Spinka and Newberry emphasised too an emotional component in learning to manage being surprised. Mishaps and shocks can disturb an animal and panicking is maladaptive. By losing control as part of play, mammals may be less likely to overreact in genuinely precarious situations. Play's execution in safe scenarios, with the thrills of surrendering control and the pleasure of recovering it, may even contribute to its elusive character of being fun.

Whatever play's exact purpose is – and it feels as if now the subject is being treated with the seriousness it deserves – its prevalence in mammals, and other animals with complex behaviour, indicates that nervous systems need careful honing and that parental care aids in providing circumstances under which this can happen. How else to explain that we humans, with brains that don't fully mature until their third decade, have childhoods that stretch nearly twenty years? It now makes more sense that much of my parenting time is spent either participating in or supervising play.

One Saturday morning – definitely more in supervisory than participatory mode – I tiredly lay on the sofa while the girls flitted from toy to toy. I was trying to read something or other, when Mariana, only just able to walk, tottered towards me. In greeting her, I felt a notion of time and development and of life's single direction. She was, I saw, heading for adulthood and her physical prime, and I was retreating from mine, and the most important thing I could do was help in any way I could to make her and her sister the best people they could be.

CHAPTER NINE

Bones, Teeth, Genes and Trees

If you look at a tree – an oak, say – you can understand that its shape resulted from its trunk having split into branches, and those branches having, in turn, split into smaller branches and so on. But to know an individual tree, it is not enough to grasp its tree-ish nature; you must see how its inherent drive to grow and branch actually manifested itself in a specific time and place. On an actual tree, there will be a scar left by the big branch lost to the storms of 1987. There will be gaps where disease has recently claimed further limbs, and no gaps where branches have grown into the space vacated by victims of an earlier malady. The wind will have shaped the tree's branches in a certain way, and the tree will be stronger on one side because of its position relative to the arc of the sun and neighbouring trees. To understand an actual tree, you must see how its biology and circumstance collided.

When I moved to Manhattan, aged 28, I went, in my first week there, to the American Museum of Natural History. I think I was lured there by the sight of a familiar face. The museum was hosting an exhibition that followed Charles Darwin from his birth to his death. The path through the exhibition was a timeline: each step forward took the visitor another year or two into Darwin's life. There was the prodigious boyhood collecting, followed by his meandering and varied education, and then the letter on which the whole story hinged: an unexpected invitation to travel as captain's companion and naturalist aboard the *Beagle*.

Beside a model of that modest vessel, a map portrayed its immense voyage, showing, surprisingly, how the fabled Galapagos Islands had occupied just five weeks of a journey of over four and a half years. After I'd perused cabinets full of specimens and drawings from these travels, the exhibition

Figure 9.1: Darwin's Notebook B: Transmutation open on page 36, showing his famous 'I think…' sketch. Source: Mario Tama/Getty Images.

described Darwin's arrival back in England, a young man seeking to settle down and establish himself. And next stood a glass case housing Notebook B in the *Transmutation of Species* series.

The leather-bound pocket-book – dated 1837–38 – was presented open. At the top of the page were the words 'I think …' and below was Darwin's tentative sketch of a diagram showing how species might diverge over time.

I've subsequently learnt that this drawing is famous, but back then I'd never seen it. I stood before it for some time. Maybe there was an irony in someone who'd eagerly swapped London for New York standing awestruck before an object from Victorian England. But there I was, transfixed by a leaf of paper, the simplest of diagrams, and Darwin's barely legible scrawl shooting off in multiple directions, exploring the implications of what he had drawn. Here was the physical manifestation of an intellect travelling to new places; the exhilaration that must have accompanied the drawing of those few lines leapt from the page.[*]

---

[*] The so-called 'I think …' sketch was not the first such branching diagram that Darwin drew, but his earlier ones were less elaborated,

When *On the Origin of Species* was published 20 years after Notebook B was written, its sole illustration was an elaboration of that sketch. On a fold-out page, as I mentioned in Chapter 2, Darwin used hypothetical organisms to chart how he thought species changed through time. In this book, the details of any particular family tree were beside the point; Darwin wanted to establish that evolution occurred and to propose a viable mechanism to explain why. Once these goals were achieved, however, *Origin* unambiguously set a challenge: if all life forms were related, the history of their relationships was mappable on a single family tree.

Over 150 years on, science uncertainly draws and redraws attempts at this single great tree of life. Contemporary renderings display a multitude of single-celled organisms you've never heard of, with only tiny branches holding the more familiar multicellular life forms. In this book, we are concerned with only the fine structure of one of those branches – the one that contains the forking mammalian lineages. The one that broke off from the mainframe 310 or so million years ago.

In private this was something Darwin himself pondered. A page torn from a notebook – probably written in the early 1850s – shows lines of mammals sprouting from 'Parents of Marsupials and Placentals', with a note emphasising how there was 'No form intermediate' between these two groups. And in 1860, within a year of *Origin*'s publication, Darwin wrote to his friend Charles Lyell from a family holiday in Eastbourne to discuss this tree. Darwin implored the great geologist to accept a single origin of mammals, urging Lyell to consider 'the whole frame, internal & external, of mammifers, & you will see why I think so strongly that all have descended from one progenitor'.

But then Darwin offered Lyell two alternative phylogenies to be weighed against one another. The primary difference lay at the root of the tree: Darwin wanted to know if all mammals

---

less annotated, and none had been prefaced quite so poetically/ charmingly/humbly.

had arisen from an ancient population of 'lowly developed' marsupials or if marsupials and placental mammals had come from 'Mammals not true marsupials not true placentals'. Both seemed equally plausible, and so, in this letter, Darwin pre-empted the ongoing challenge of phylogenetics: deciding between different possible versions of history.

If a tree is to be used as a metaphor for a phylogeny, its buds or leaves are akin to living species, while its hardwood lines represent deceased predecessors.* The buds, the living species, are available to us now. To classify them genealogically we infer from their degrees of similarity how closely related they probably are. A chimpanzee and a human are clearly near relatives, but a chimp and a bonobo are even more alike and so must share a more recent common ancestor.

Then, from the living, we can speculate about the nature of these long-gone shared ancestors – for lions, tigers and domestic cats, say, we picture something recognisably feline, but what about the last shared ancestor of these and wolves? What we might truly know of these long-dead animals lies with the infinitesimal fractions of them that have ended up preserved in rock. 'Fossils,' wrote George Gaylord Simpson in 1945, 'are documents that free us from the limitation of studying history only by its results at one given time, a time purely accidental from the standpoint of phylogeny as a whole.'

In *The Principles of Classification and a Classification of Mammals*, Simpson said he thought this was 'one of the worst times' to study mammalian relationships because of the 'considerable gaps' between living types. Fossils provide glimpses of the ancient animals whose lives spanned these chasms. And, on occasion, they do a remarkable job. *Ambulocetus* – Latin for 'walking whale' – is one such find. A series of fossils now charts the transformation of terrestrial tetrapods into gracious sea creatures, making the assertion that whales are close relatives of hippos and cows much easier to stomach.

---

* Another notebook of Darwin's reveals that he wondered if a coral, where the basal branches are dead, might make for a better metaphor.

Here, we'll take mammalian phylogeny in the three parts discussed in the Introduction. First, the pre-mammalian period – the 100 million years that followed the divergence of mammalian and reptilian ancestry 310 million years ago. To trace the shifting morphology of the bones left behind in this period is to see mammals emerge from their reptile-like beginnings. We'll take various bony elements of the body and consider how each of them changed until we reach the creatures born around 210 million years ago that were the first true mammals.

The second period – running from the birth of mammals to the moment, 66 million years ago, when a meteor put an end to the reign of the dinosaurs – considers the first two-thirds of mammalian history, when fur marked an animal as a member of a zoological underclass. But, as more and more mammalian fossils are unearthed from this time, the story of our apparently oppressed ancestors becomes ever richer.

Third, and finally, the post-dinosaurian epoch. The mammals that awoke the day after that astral rock struck the earth lived in a profoundly new world; a world that would witness an outrageous proliferation and diversification of mammals, a thorough exploration of the possibilities of mammalian-ness. And it is here that one species of mammals tries to assemble a family tree that accounts for the origins of all 5,000-plus of its nearest and dearest.

The first two are stories written almost exclusively in the shape of petrified teeth and bones. The third begins with fossils, but to decipher relationships between living mammals there's much more data to play with. Simpson's view, in 1945, was that 'morphological data and paleontological data (also mostly, but not exclusively, morphological) always have been and (barring some wholly unheralded and most improbable achievement in some other field) always will be the principal basis for the study of phylogeny.'

He added, however, that four other sources of data – genetics, physiology, embryology and geography – also had roles to play. On the latter, he reminded readers that 'Animals clearly cannot have common ancestry without also having common geographic origin.' Where animals live and where we find fossils matters. Two similar species found in adjacent territories are more likely closely related than two on opposite sides of the globe. 'Some classifiers,' he wrote, 'deny that geography has any useful bearing in phylogeny [...] but we need not take them seriously.'

While we'll see how prescient this remark was, something that had seemed unheralded and improbable in 1945 did indeed soon happen in another field: as we saw in Chapter 3, DNA sequences provided an additional means of tracing historical relationships.

To the first true mammals

I have taken as my definition of a mammal the most widely used one: the one that says a mammal has a dentary–squamosal jaw joint. Whereas in other amniotes two different bones form the jaw joint. Honestly, bones and fossils are difficult. This essentially means that we mammals have lower jaws that consist of a single bone – the dentary, repeated on each side – which holds all our bottom teeth and directly forms a joint with a skull bone, the squamosal, which is usually fused to other skull bones. This union forms the pivot point of our jaws. In humans, it's just in front of our ears.

In addition to their unfamiliar names, I think I've always been put off bones by an ingrained association between them and death. The skeleton hanging in an anatomy classroom feels like the antithesis of life, doesn't it? A skeleton's relationship with the dynamic process that is life has always felt weak, a passive scaffolding around which the interesting stuff happened.

Today, I'm not saying I'm a fully signed-up bone aficionado, but I'm more of one than before. Take that jaw joint again. True, it's a convenient, commonly fossilised anatomical landmark that allows palaeontologists to classify fossils as mammalian, pre-mammalian or non-mammalian. But it's also really important. The dentary–squamosal jaw joint helps to equip mammals with strong jaws that are capable of sophisticated chewing motions.

Mammals becoming warm-blooded generated increasingly immense energy demands – a mammal must eat about 10 times more food than an equivalent-sized reptile – so, to fuel such a lifestyle, calories must be liberated from food quickly, efficiently and as completely as possible. And that starts with chewing – mastication is elemental to mammalian biology. The jaw architecture of a fossil mammal or protomammal, rather than being a dry taxonomic marker, gives an idea of how well that animal chewed its food, how proficient a calorie gatherer it was, how badly it needed to release energy quickly.

The thrill I had in New York, standing before Notebook B – feeling as if I had sensed a fire of creativity that had burnt 180 years before – is akin to what the trained eye can derive from a fossil. Fragments of bones and teeth evoke, if you understand them well enough, an animal in life. Warm- or cold-blooded, insectivore or herbivore, tree climber or burrower, all this can be known from left-behind hard parts of a body. The skeleton is not a passive scaffold; it shapes and is shaped by its possessor's life.

In eyeing the 100 million years that led to the dawn of mammals, the goal is not then to drily discuss the names of bones that form various joints, but to recover an idea of the types of animals to which these bones belonged. What, for a living animal, does the articulation of its limbs mean? What does the distribution of a long-dead animal's ribs say about how it moved and breathed?

Tom Kemp, a palaeontologist at Oxford who has dedicated his career to the protomammals that populated this epoch, arranges these animals in successive 'grades' in his book *The Origin and Evolution of Mammals*. The first grade is barely distinguishable from the reptilian ancestors it lived alongside, but each subsequent grade is more mammal-like. At the broadest level, there are three main pre-mammal tiers (although Kemp believes it's possible to define as many as 10 grades): simply put, first came the pelycosaurs, then the therapsids, and then the cynodonts. In the murky region between cynodonts and true mammals laid the mammaliaforms.

What is important to grasp about these successive grades is that we're not talking about a single thin lineage that progressed from reptile-like beginnings to a mammalian destiny. As we touched upon in Chapter 2, each grade was associated with what's called a 'radiation' (see Figure 3.2). When we talk about pelycosaurs or therapsids, we talk about animals that share various core features, as mammals do today, and which came in a variety of forms – small and large, herbivores, carnivores, insectivores and omnivores – as mammals do today. Interestingly, Kemp sees the fossils as indicating that each new grade began with a small carnivore that had made the essential steps forward. Each time, diversity radiated out from a single point, and each successive radiation eventually – over tens of millions of years – replaced the one that had come before it. The implication is that a set of steps towards mammalian biology allowed a new type of animal to radiate into many species that outcompeted members of the older grade.

To consider these grades, we'll divide up their ancient skeletons by moving from front to back – from teeth, back through the jaws and the rest of the skull. Then, beyond the head, the focus will be on spines, limbs and ribs. All of these structures evolved in ways that say something about the types of animal that sequentially came into existence.

Teeth

Teeth are a big deal in mammalian palaeontology. First of all, mammalian teeth are incredibly durable objects. Their enamel coating is the hardest, most mineralised substance in a mammalian body, meaning teeth are the most commonly fossilised and best-preserved parts of ancient mammals. Second, teeth are incredibly informative. By looking at just a single tooth, an expert can tell what a mammal ate, therefore grasping its general lifestyle, and also get an idea of how big the mammal was.[*]

---

[*] I have sometimes wondered whether people overreach in these inferences. You will, occasionally, encounter a rather elaborate description of a mammal that lived, say, 150 million years ago, then at the end learn that this is known only from a single fragment of a

And third, perhaps most fundamentally, mammals, more than any other group of animals, have diversified their dentition to make their mouths not just hunting devices but vital contributors to the digestive process.

I'm not sure many modern humans intuitively grasp the centrality of teeth in animal life. We very rarely interact with the wider world using our teeth or mouths; seldom do we acquire food directly with them either. We have hands, and we have cutlery. But think of almost any other animal, and the first contact – often the *only* contact – it makes with its food is via its mouth. Sure, other primates use hands, and certain rodents employ their front paws – while certain birds and carnivores hunt with talons and claws – but mostly the mouth and its teeth constitute an animal's primary means of gathering food.

This brings me to a confession. After starting to think about all of this, I sat at my desk, peeled an orange, broke it apart onto a plate – procedures admittedly executed by my nimble fingers – then ate a segment without using my hands. I felt self-conscious moving in on it, and with my mouth inches from the plate, I felt completely ridiculous. But when I put my teeth around the segment, things got interesting. I picked up the orange with my incisors – the spade-like teeth at the front, which gather and bite food. I then transferred the segment to the rear of my mouth and chewed it – grinding and slicing the fruit using the elaborate crowns of my molars. As I did so, my snack was infused with the digestive enzymes in my saliva. Occasionally, my tongue flicked the segment forwards in my mouth and again sliced it with my front teeth. Only after some time was it ready to be swallowed. Orange segments, being a rather inert foodstuff, hadn't needed to be immobilised by a swift stabbing from my (admittedly limited) canines – the teeth that flank my incisors and that are elongated in cats, dogs, other mammalian carnivores and vampires). I was so pleasantly

---

molar. Such extrapolation seems standard in palaeontology. Its origins appear to be a 1798 assertion by Georges Cuvier that tooth structure correlates with all the other organ systems of an animal, and that from a single tooth he could reconstruct a whole animal.

surprised by the naturalness and efficiency of the whole experience I ate a second segment the same way.

This division of dental labour is the key to understanding mammalian teeth. Those three main categories – incisors, canines and cheek teeth (divided into molars and premolars) – are specialised to achieve different goals. The cheek teeth are especially elaborate; with these, mammals have added extensive mastication to the ancestral tooth function of simply biting.

Vertebrate teeth evolved originally in fish, many of which have thousands of teeth throughout their mouths and into their throats. Amphibians have fewer teeth but still a lot, reptiles have fewer still and – talking at these levels of generality – mammals have the fewest teeth. Paralleling that reduction in numbers, teeth became increasingly restricted to two complementary upper and lower arcs. As a rule, reptile teeth tend to be repeating arrangements of the same peg-like structure, and this pattern of many similar teeth is what the first terrestrial amniotes had.*

To progress along the pre-mammalian lineage, dimetrodons – the sail-backed pelycosaurs from 295 to 270 million years ago that are often mistaken for dinosaurs – are a good place to start. The clue is in the name: 'dimetro-' means 'two measures' and '-don' refers to teeth. However, their dentition suggests they should have been christened *tri*metrodons. These animals had large incisor-like teeth, flanked by pronounced stabbing canines, and another type of sharp, serrated and curved teeth behind those. Respectively, these teeth would have grasped, disabled and then securely retained prey. Another species of sail-backed pelycosaur had smaller peg-like teeth that were

---

* I shouldn't completely dismiss non-mammalian teeth. Over 300 million years, certain reptilian lineages *have* evolved mouths containing more than one sort of tooth, including some living lizards. In birds, however, teeth have disappeared altogether. Instead, birds evolved gizzards – muscular parts to the stomach that, when lined with grit, grind down food, which is highly efficient. Teeth are just one way of speeding your digestion.

much better for grinding tough plant material, and that species was therefore almost certainly a herbivore.

Examining the dentition of animals progressively closer to true mammals shows that the division of labour between teeth became more and more pronounced. In the next grade, the therapsids, differences between the three types of teeth are even more pronounced, and in early cynodonts the dentition is recognisably mammalian. In particular, the teeth behind the canines have become much more complex, especially those furthest back – the prototype molars – which have evolved single primary cusps flanked by accessory ones. Peering into the mouths of these 255-million-year-old animals, it appears that these cheek teeth had become the main site of food manipulation.

Many therapsids probably feasted on insects, catching them between their front teeth, just as many insectivorous mammals still hunt (a much trickier proposition than picking up a segment of orange). But while insects are an excellent source of nutrition, the good stuff lies under their tough exoskeletons. The evolution of crushing and grinding back teeth would have been a boon for cracking open this brittle coating and freeing the innards.

In cynodonts, the cusped cheek teeth didn't yet perfectly interlock as they do in today's mammals, but they were getting there. Whether it's the biting front teeth or the chewing back ones, occlusion – a precise alignment between the top and bottom teeth – transforms dental efficiency. 'Imagine trying to cut with scissors,' writes Peter Ungar, a tooth expert at the University of Arkansas, 'with blades that don't line up.' For both biting and chewing, a precise alignment between upper and lower teeth boosts effectiveness hugely.

In addition to the evolving jaws, which we will discuss next, occlusion was facilitated by a reduction in the number of times teeth were replaced in early mammals. When the mammalian lineage came to grow only milk and adult teeth, this allowed upper and lower teeth, which would be partners for life, to develop simultaneously and to match each other perfectly.

People who work on fossilised teeth study the scratch marks on them very closely to see how they moved against each other as an animal ate. What these traces indicate is that it's no good talking about teeth in isolation. To gather food, and nimbly chew it to pieces, teeth needed to reside in jaws capable of using them appropriately.

Jaws

If you gently press a finger to your temple as you clench your teeth together, you can feel the flexing of a muscle. Then, if you move your finger down just past your ear and repeat, you feel another large muscle contract. Together, these muscles give contemporary mammals very fine control of their jaw movements. The two of them pull in different directions so that we can not only open and close our mouths but can also move our lower jaws from side to side. This latter movement was the key to mammalian chewing, and it took some 100 million years to evolve.

The lower jaws of pelycosaurs consisted of three bones running front to back. There was a single muscle on the inside of the jaw and a single one on the outside, and the lower jaw moved only up and down. These muscles ran from attachment points on the skull to the very back of the lower jaw, meaning the bite force wasn't great. Imagine holding chopsticks. If you move them from their far ends – as these muscles moved the jaw – the force you can generate between the food-grabbing ends is limited. But if you hold the chopsticks halfway down, you can press the tips together much more firmly.

The trends we see in the pelycosaur–therapsid–cynodont progression are, first, greater bite strength, then more finessed movements. The muscles on the outside of the jaw expanded and became more of a sling around the lower jaw – that is, they moved from being solely at the back of the chopstick towards pulling it from the middle. The largest lower jaw bone – the dentary – first developed a protrusion for better muscle attachment and then enlarged all over. A second group of muscles running to the outside of the jaw began to form. The ones you feel when you touch your temples were always there, but the lower ones, the masseter muscles, were a mammalian innovation – they appeared first in therapsids, and became functionally important in cynodonts.

The expansion of the dentary bone and the enlarged muscles attached to it both made for a stronger bite, and importantly the shifting geometry of the jaw transferred tension away from the jaw joint, exerting greater forces through the teeth, where the power was more useful. So, the more protomammals evolved, the less you would want to be bitten by one.

The gradual evolution of the masseter first helped stabilise the jaw, but then this muscle changed further, enabling precisely controlled sideways movements of the jaw, and therefore new modes of chewing. Strong but finely tuned jaw motions would have permitted the increasingly sophisticated cheek teeth of cynodonts to make these animals' mouths ever more efficient at kick-starting calorie liberation from whatever they were eating.

To transition fully to mammals, we need to return to the dentary. After all, this bone becoming part of the jaw joint is what defines a mammal. The dentary was always the main teeth-holding bone of the lower jaw, but, initially, when it was only one of a number of bones in that jaw, it was a rather slender one at the front. As the protomammalian lineage evolved, and more muscle attached to the dentary, this bone became deeper and started to extend further back (see Figure 9.2). Consequently, the bones behind it – one of which was forming the original jaw joint – got smaller and smaller as that jaw joint became less important, owing to the new muscles' orientations placing less stress on it. Eventually, the dentary directly joined to a skull bone to form the jaw joint that is unique to mammals.

For a long time, the loss of a jaw joint and its replacement by a new one in which the dentary met a different skull bone, the squamosal, seemed a most unlikely transformation. Understanding how increasingly small amounts of tension had acted on the ancestral joint helped explain its diminishing

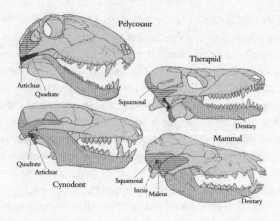

Figure 9.2: The advent of true mammals was associated with a new jaw joint evolving.

mechanical importance and the possibility of change. Nevertheless, the discovery of *Probainognathus* in Argentina in 1970 represents one of the most elegant intermediate forms ever dug up. This fossilised cynodont had two jaw joints – a new mammalian type joint sitting beside an older reptilian type one.

The new joint may have permitted a greater range of motions, offering a new pivot point for muscular forces involved in side-to-side chewing. However, another selective force may have helped drive its evolution. After the new joint was established the small jaw bones behind the dentary went off to enjoy a most wondrous second career as parts of the mammalian middle ear – a story we'll pick up in Chapter 11. However, vibrations in these bones may have contributed to hearing before they'd ceased being part of the jaw. Hence, the jaw's reconfiguration and the middle ear's emergence were tightly linked. Selection for both improved mastication and greater hearing acuity – particular at high frequencies – would together have shaped morphological change.

Nose

Moving nosewards, there is a final oral innovation to note. In you, me and all other living mammals, the nasal cavity and mouth are separate spaces, conjoined only at their rear ends. We can breathe through each, but the nose also sniffs and the mouth eats. That, however, is not how things started out; once there was just a single cavity – an arrangement that persists in most reptiles. Mammalian ancestors evolved a second bony palate. Akin to inserting a mezzanine floor in a large open building, outgrowths from the two sides of the upper jaw extend to fuse centrally and to create separate mouth and nasal cavities.*

Why secondary palates evolved, however, is a matter of debate. The issue is that they have multiple useful functions. First, they strengthen the upper jaw, which may have been the original driving force behind their evolution, owing to the increasing forces being exerted on the jaw. But then a secondary palate also allows mammals to eat and breathe at the same time. To avoid choking, we mammals need only to stop breathing while we swallow, which means we can simultaneously devour food and oxygen, increasing the rate at which these twin fuels of our warm-blooded existence are consumed.

The secondary palate is also elemental to suckling. With it, young mammals form a vacuum around their mothers' teats to draw in milk. They could do that without a secondary palate, but they need one to eat using a vacuum system and to simultaneously breathe.

Above the palate, the nasal cavity – which is large in mammals – functions both to smell and to breathe. And, in mammals, these processes both take advantage of arrays of scrolled bones called turbinates lying in the cavity. In each case, these folded sheets of bones serve to expand the surface area of the nasal cavity. Turbinates that facilitate odour detection are covered in the sensory cells that bind airborne chemicals: more bone equals more sensory cells, which equals greater odour perception. The respiratory ones are coated in mucous-secreting cells and lie in the main route of airflow, where they act as a sort of natural air conditioner.

---

* It strikes me, therefore, that compartmentalisation of multipurpose spaces is a feature of mammalian evolution at both ends of the body.

Air entering a nose tends to be dirty, cold and dry. But, as it passes over and around the respiratory turbinates, it is warmed, dirt sticks in the mucus, and it is moistened by water evaporating into it. All of which – particularly the moistening function – makes the air less shocking to the lungs. In addition, when air is exhaled, the water previously infused into it also condenses back onto the turbinates, instead of being expired and lost.

This process has long been recognised but was traditionally thought to represent a water-saving adaptation to living in dry environments. In the early 1990s, however, Willem Hillenius, at Oregon State University, made the case that respiratory turbinates evolved in response to the huge ventilation rates necessary for being warm-blooded, which could otherwise lead to colossal water loss. Indeed, the similarly warm-blooded birds have elaborate nasal turbinates too. Hillenius and colleagues then suggested that determining when turbinates evolved would tell us when the mammalian lineage became warm-blooded. Their initial claims for therapsid turbinates were a little shaky, but subsequent finds seem to support the idea that these protomammals did, indeed, have nasal water-saving devices.

Cranium

Above the nose lies the brain-housing cranium. Brains are the subject of Chapter 12, so I will say very little about them here. The one simple thing that *is* worth reporting is that mammal brains are big, and they appear to have become big only when true mammals evolved. Before that, cynodont brains might have been somewhat enlarged relative to their ancestors, however it's hard to tell because these brains were encased not in bone, but in cartilage braincases within the cranium.

Beyond the skull

The fossilised 'postcranial' body parts of protomammals hold two main messages for us. First, there are further important indications about how these animals breathed, and second, we see how they moved. Mammals developing a mechanism that allowed them to eat and breathe simultaneously was important, but equally influential was their evolving a way to *run* and breathe at the same time.

When tetrapods arrived on dry land, with their legs freshly evolved from fins, these limbs sprawled out to their sides, and they ran using a hangover method from their fishy ancestors. Their bodies moved in lateral waves which helped swing their limbs forward. Lizards still run like this, the arcs of their hips contributing to their stride length. This sort of lateral body bending presents a problem, however – on sequential strides, the left and right lungs are alternately compressed. An animal cannot breathe while running this way – air just passes from the squashed lung to the uncompressed one on the other side, then back again on the next stride.

Interestingly, this lateral flexing was lost early. Looking at how pelycosaur spinal columns locked together indicates that they could no longer bend side to side. They couldn't bend from front to back either (as we do when we bend to touch our toes) – this came later – but it says these animals were moving in a new way, probably able to breathe as they did so.

When mammals later added an ability to flex their lower spines front to back, it enabled a new kind of fast, bounding gait. And running this way actually aided breathing by compressing and releasing the two lungs in unison.

The second note on breathing concerns a uniquely mammalian characteristic that doesn't fossilise. Only mammals have a muscular diaphragm spanning the entire torso at the base of the chest cavity. This sheet of muscle works with rib muscles to fill the lungs. As the ribs move up, the diaphragm moves down to further expand the chest cavity and draw in air more powerfully.

When the diaphragm first evolved can't be known directly, but perhaps the ribs are telling. Pelycosaur and early therapsid ribs ran all the way back to their pelvises, but late therapsids' ribcages stopped at the end of the chest cavity, suggesting these animals possessed this useful sheet of muscle.

To understand the ways in which tetrapods move it is important to appreciate that a tetrapod is like a wheelbarrow. The essential similarity is that a wheelbarrow's front wheel, like a tetrapod's front legs, doesn't power the wheelbarrow's forward movement. All the thrust comes from the legs at the back. The primary purpose of the front legs – like the wheel – is to keep the body off the ground as they carry it forward.

Beyond changes to the vertebral column, the main story of how mammalian locomotion evolved relates to how two pairs of ancestral, lizard-like legs that sprawled out to the sides got tucked underneath the body. The front legs became more mobile for guiding the animal forward, and the back ones found new angles at which to power this movement.

This involved many incremental changes to numerous joints, limb bones and their attached musculature. And a great many mechanical inferences about animal movement can be drawn from fossilised shoulders, hips and legs. Drawing on progressively younger fossils, Tom Kemp has reconstructed the likely sequence of events that mediated this transition.

Pelycosaurs may have changed the spinal column, but not much of note happened with their heavy, sprawled legs. The front legs were attached to massive shoulder girdles that were themselves firmly bound to the ribcage. The hind legs had equally restricted movement. The early therapsids that followed the pelycosaurs, Kemp writes, would have been 'doubtless slow and clumsy by modern standards', but some fundamental changes had begun to occur. At the front, the legs still protruded to the sides, but the shoulder girdle was 'less massive' and freer to move, and the nature of the shoulder joint was modified, making the forelimbs much more mobile. These limbs may not thrust an animal forward, but their manoeuvrability is elemental to an animal's overall mobility.

And at the rear ends of therapsids, a pretty radical transformation was taking shape. In 1978, Kemp presented his 'dual gait' hypothesis, proposing that when therapsids didn't need to move quickly, they walked much as their ancestors had, with four legs splayed to the sides. But when they needed to move sharpish, therapsid hindlimbs moved underneath the body, and with the knees turned forward, they ran much more as mammal legs do. A similar duality is seen in crocodiles and iguanas today. The idea of upright hindlimbs with sprawling front ones certainly strengthens the wheelbarrow image.

Among the cynodonts that followed therapsids, progress was initially slow, but eventually the hindlimbs became permanently arranged upright, below the body, and, in later cynodonts the front legs also transitioned to such a position. The shoulder joints and pelvis also became entirely mammal-like. Kemp sees animals that were 'increasingly able to accelerate and change direction'.

An animal emerges

At the start of this book, I said one potential way to arrange a series of chapters about various mammalian attributes had been to order them according to when the respective traits had evolved. The above shows most clearly why this was untenable: for 100 million years eating, running and breathing mechanisms all evolved in parallel. Various definitive mammalian traits inched into existence side by side.

Perhaps the co-evolution of different elements of a larger system is most evident when it comes to eating. The emergence of complex cheek teeth capable of grinding down tough food could easily be viewed as a breakthrough event. But those teeth would have been of no use – in fact they probably wouldn't have come to exist – if complementary changes in the jaw bones and musculature hadn't accompanied their evolution.

The emergence of a strong, mobile jaw toothed this way was part of a trend towards protomammals becoming increasingly energy-demanding animals. Food was caught more effectively and consumed more efficiently, animals didn't have to stop breathing while chewing, enlarged nasal cavities were equipped with scrolled bones that prevented water loss in warm, expired breath, the animals could run and breathe, and diaphragms helped pump in and out greater lungfuls of air. All of this points to active animals driven by faster metabolisms and eventually piloted by larger brains.

The posture and gait of the progressively more mammal-like grades also present an animal that seems faster and more dynamic. For a long time, the upright posture of mammals was viewed as superior to reptiles' low stance, but detailed analyses of reptilian movement have shown that it can be similarly fast and effective. Kemp has suggested that the selective forces that sculpted mammalian legs might have made these animals more agile and manoeuvrable, so as to cope better with difficult terrains.

Finally, it is interesting to look at where these ancestors lived. Rock sediments don't just harbour left-behind teeth and bones; they also indicate the climates in which the rocks formed. Pelycosaurs lived around the equator of a world in which all land masses were joined in a supercontinent that we call Pangaea. There, conditions were permanently warm and humid, a boon for animals still – geologically speaking – relatively fresh out of water.

In contrast, therapsids evolved away from the equator, in conditions that were cooler and more seasonal. These animals must have been resilient to air that was both cold and dry. Then, having evolved in a school of hard knocks, therapsids later returned to equatorial regions and succeeded the pelycosaurs.

However, conditions don't just vary according to latitude and season; they change with the earth's daily rotation. The ability to tolerate an absence of sunshine also aided early mammals moving into another ecological niche – the night, a necessary tactic for avoiding their domineering reptilian contemporaries.

Living with dinosaurs

When pelycosaurs first evolved they quickly became the dominant terrestrial vertebrates of their day. Likewise, therapsids soon radiated widely and successfully. But then something happened: 252 million years ago, the grimmest mass extinction in Earth's history wiped out up to 95 per cent of marine species, vast swathes of insects and maybe two-thirds of terrestrial vertebrates. No one knows for sure what caused the End-Permian extinction, but massive and sustained volcanic eruptions stoked widespread fires and the chemical equilibria of the atmosphere and the oceans collapsed. Global warming rampaged, and entire ecosystems crashed. The post-apocalyptic fossil record suggests biodiversity in the new Mesozoic period took 10 million years to recover.

Whether the survival of the particular therapsids and cynodonts that made it through this cataclysm can be attributed to their increasingly mammalian physiology is moot. Certain energy-grabbing skills, for example, may have been useful, and, interestingly, by far the most common amniote in the extinction's aftermath was a burrowing, pig-like therapsid called *Lystrosaurus*. Somewhere alongside these animals, however, lived the survivors that would evolve into dinosaurs. In fact, there was quite the evolutionary jamboree back then, with the lineages that would lead to modern lizards, frogs, turtles and crocodiles all beginning.

The world had been reconfigured. And the dinosaurs emerged from the chaos as the dominant land animals. Unlike pelycosaurs and therapsids, when true mammals evolved 210 million years ago they had about 145 million years to wait before they formed dry land's foremost fauna.

Although the goal of this book is to ask what defines a mammal and how these things shape the way modern humans live, I acknowledge that I occasionally get swept up in a celebratory tone and narrative that might convey the idea that a mammal is a superior animal. But, truly, this business of mammals spending the first two-thirds of their existence playing second fiddle to dinosaurs is as good a reminder as anything that this is not the case. I rate mammals very highly, but there is more than one way to make a living in this world, and, for a long time, dinosaurian biology, we must concede, trumped mammalian biology.

What, then, did mammals get up to for these 145 million years? Until very recently, the general view was: not a lot. Mammals were seen as having held on to their simple existences by eating insects in the dead of night, so as to avoid the reptilian overlords. A rather limited fossil record suggested a rather limited set of animals. But over the last 20 or so years, this worldview has been obliterated.

Finding mammals of the right vintage has always been about finding the right rock deposits. For a long time, palaeontologists have travelled to the Gobi Desert for them. Recently, however, things got more interesting in explorations in Greenland – at Arctic sites made risky by polar bears – South America, and most impressively in rocks laid down in today's north-eastern China by periodic volcanic eruptions around 160 million years ago.

The sites in Greenland and China have provided a mass of mammals whose skeletons indicate they occupied a great many ecological niches. One fossil's tail bones show a beaver-like tail, its hands carried webbing, and its teeth were made for catching fish. Another had a snout that would have sucked up ants; a compatriot climbed trees; yet another had membranes between the front and rear legs evoking a gliding existence akin to today's flying squirrels. Insects were definitely a crucial food source, but diets were diverse. Among mammalogists' favourite finds is a mammal that lived around 125 million years ago, who died while digesting a small dinosaur. There may be no evidence that mammals in the Mesozoic were ever larger than a badger or small dog, but they were a fantastically mixed bunch doing most things that small mammals do today.

The now thousands of Mesozoic mammalian remains – sometimes whole skeletons, more often jaw fragments and teeth – were recently all compared to one another, to construct a Mesozoic phylogenetic tree and to infer the speed at which mammals were then evolving. A team led by Roger Close, then at Oxford University, showed that a burst of morphological changes occurred among mammals in the mid-Jurassic period, 180 to 160 million years ago.

Remarkably, mammalian bodies were changing at rates 10 times quicker than they were at the end of the dinosaurs' reign. Close said, 'We don't know what instigated this evolutionary burst. It could be due to environmental change, or perhaps mammals had acquired a "critical mass" of "key innovations" – such as live birth, hot-bloodedness, and fur – that enabled them to thrive in different habitats and diversify ecologically.' This creative period coincides with the earliest known therian mammals, the ancestors of today's placental and marsupial mammals. But the rocks also reveal that many different mammalian lineages arose around this time, most of which remained buried in the Mesozoic, although one rodent-like line survived until 30 million years ago. Evolution, it is clear, relentlessly experiments.

One feature of therian mammals that may have aided their flourishing was the evolution of another new iteration of the molar tooth. The 'tribosphenic molar' was born at the base of the therian radiation; the cusps of this tooth shear at one end and grind at the other. It is a dental masterpiece. The tribosphenic molar may have been particularly good for taking advantage of the seeds and fruits of the newly evolved flowering plants that began to flourish in the Jurassic. Early mammals also lived through the break-up of the old supercontinent Pangaea, as Earth's land masses gradually began to resemble today's continental arrangement.

Dioramas in natural history museums that feature Mesozoic mammals show them as savvy little animals among the dinosaurs; the overall effect is of a world that is strikingly and thrillingly alien. Conversely, dioramas that imagine scenes from the last 66 million years are at first glance familiar. Their strangeness emerges only with closer inspection – the head of the rhino-like beast is odd, the predator's teeth are unusually long, the horses are too small, the armadillos too big, the horns of the antelopes too bizarre. Foreignness lies in the details, while often the basic types of mammals appear familiar.

These types we recognise correspond to the 17 or so orders into which today's 5,000 or so placental mammals are classified, each order reflecting a group of mammals that live a certain way. There are the gnawing rodents whose incisors never stop growing, the flying bats, the primates, through the carnivores and the insectivores, down to the smallest orders that respectively contain only pangolins, aardvarks or the flying lemur. As we move toward the present, the question posed is whether diminutive predecessors of these living orders existed in dinosaurian times or whether modern groups only emerged after the fateful meteor killed off the dinosaurs.* This cataclysmic event, known as the KT (Cretaceous–Tertiary) boundary, was a giant event in mammalian history, but the exact nature of the zoological creativity that produced today's mammals remains surprisingly enigmatic.

The living

If Darwin's letter to Charles Lyell questioning the long-gone root of the mammalian phylogeny foresaw how biologists would always have to choose between alternative historical scenarios, the first attempts to systematically infer the relationships of living species based on their morphology abundantly confirmed this issue.

The first such phylogeny was published by St George Mivart – yes, that St George Mivart, the doubter of evolution's ability to craft a mammary gland – who, inspired by Darwin, had set to work trying to determine the interrelationships of primates. First, he did so using similarities and differences between their spinal columns, publishing in 1865 a family tree of 29 different species, including humans.

But then Mivart published the second ever phylogeny. Again, it charted primates, but this time he inferred relationships from their limbs, and the resultant tree was entirely different from the first.

Immediately, the difficulties inherent in recreating history by comparing living species was laid bare. How could it be

---

* Bar, of course, those that survived as birds.

determined whether vertebral columns or limbs were better placed to reveal true ancestry? And how could different data sets be combined? The proposed solution was always more data and more analysis. No single trait – whether it was spines, limbs or placentas – was going to be sufficient to arrange the morass of living mammals according to their genealogy. Many aspects of living and extinct species – their morphology, development, distribution, physiology and genetics – were required before weighing up which might be more closely related to which.

But as Simpson discussed in his seminal 1945 classification of mammals, a major problem was convergent evolution. The marsupial and placental mammal radiations had, for example, generated some very similar animals. Thylacines – the so-called Tasmanian wolves (or tigers) – had many more characteristics in common with the placental wolf than they did with kangaroos. However, the traits thylacines did share with the kangaroo – such as their mode of reproduction – were 'more basic, or important, or essential', and, therefore, accorded much greater weight.

The classification that Simpson produced was the benchmark for mammalian phylogeny for the next half-century. He arranged living mammals in 18 orders – one for monotremes, another for marsupials, and he split eutherians into 16. Orders are generally pretty uncontroversial and obvious groupings, such as primates, rodents, and proboscideans (elephants and their extinct relatives). The greater challenge has always been to say how these orders are linked. To this end, Simpson produced four unevenly sized eutherian 'cohorts'. One included only the cetaceans, one combined rodents with rabbits and their near relatives; we humans were with our primate cousins in a large third group that also contained bats as well as insectivores and the South American sloths, anteaters and armadillos, plus the Asian pangolins, or scaly anteaters. The fourth group housed everything else, including ungulates (hoofed mammals) and carnivores (cats, bears, wolves and seals), which fossil finds seemed to suggest were closely related.

From the 1950s onwards, phylogenetics got much more statistical, and periodically minor to medium tweaks were suggested to Simpson's classification, but, in general, it stood strong. When Michael Novacek published another landmark phylogeny in 1992, it wasn't hugely different from Simpson's. Alongside this family tree, Novacek's article – entitled 'Mammalian phylogeny: shaking the tree' – surveyed the various ways in which people were trying to employ new techniques to verify or challenge it. Pretty soon, after certain molecular investigations had suggested a guinea pig wasn't a rodent and that monotremes had branched off from marsupials *after* marsupials had branched off from placentals, Novacek treated studies of protein and DNA sequences equivocally.

In 1997, though, genetic data didn't so much shake the tree as forcibly rearrange the main branches. A new study (playfully) entitled 'Endemic African mammals shake the phylogenetic tree' was published in *Nature*. This study ripped apart two groups previously established on shared morphology and created a new group based on geography.

Previously, telling similarities between the superficially distinct elephants, aardvark, manatees and dugongs had meant they were grouped together, and more recently they'd been linked to the enigmatic hyraxes – stocky little herbivores that look as if someone has stuck a squirrel's head and four short legs on a rabbit's body. But suddenly they had company.

Elephant shrews, so-called for their long noses – they're not gigantic shrews – had always been considered insectivores and grouped with proper shrews, hedgehogs and moles, whereas golden moles and tenrecs had typically been placed with rodents. In extensively analysing the DNA sequences of five separate genes across many mammals, Mark Springer and his colleagues at the University of California Riverside, showed that elephant shrews, golden moles and tenrecs were all most closely related to elephants, aardvarks, manatees and dugongs. What linked them was Africa.

According to the genetic data, this group of wildly different mammals represented a single mammalian radiation stemming from a single ancestor. Africa had been cut off from other continents for millions of years, and these lineages had evolved to fill many of its ecological niches. Not all possible types of mammals are there, but many are. This echoes the convergent forms seen between marsupial and eutherian mammals, as the African group yielded animals that looked much like mammals that had evolved independently on other land masses.

Then in 2001, Springer's group revealed an even more comprehensive overhaul of mammalian phylogeny. It was published alongside work from Stephen O'Brien's team at the US National Cancer Institute, which had reached identical conclusions. Examining reams of genetic data – so much that it had taken the best part of a decade to pull it together – these two teams said placental mammals were best described as four separate radiations:

1 Afrotheria, the new name for the group originating in Africa.
2 Xenarthra, comprising the toothless South American sloths, anteaters and armadillos.
3 Laurasiatheria, named after Laurasia, a supercontinent which included today's North America, Greenland, Europe and most of Asia, where the ancestors of insectivores, carnivores, ungulates, cetaceans, scaly anteaters and bats had evolved. Bats being torn away from the primates was another big surprise. These two groups had always been believed to be close cousins.
4 Euarchontoglires, which contains humans. This radiation consists of primates and some close relatives, along with rodents and lagomorphs (the rabbits and kin).[*]

The advantage DNA has over morphology is that it produces masses of unambiguous data. Every base in a sequence of thousands is definitively an A, C, G or T. Then, for comparing species, every A, C, G, and T acts as an individual trait.

---

[*] Recalling this book's origins, it seems that testicles were externalised in placental mammals when the stem of groups 3 and 4 separated from the ascrotal Afrotheria and Xenarthra.

Additionally, much of this data is free of certain caveats of morphology; for instance, genetic change can be neutral in its effects on phenotype, so invisible to natural selection. And, perhaps more importantly, while selection can cause animal forms to converge, the chances that they do so via the same genetic changes are infinitesimal.

These two papers turned mammalian phylogeny on its head. In digging into the genetics of living mammals, geography had suddenly been thrust to the fore, which makes perfect sense. And the results again spoke to how mammals left to their own devices independently converge on similar forms.

This tree does, though, still have people scratching their heads − for example, no single morphological marker that categorically marks Afrotherians as Afrotherians has been found. Plus, genetics hasn't cleared up everything. Among remaining questions, uncertainty persists as to the order in which the lineages leading to today's four groups diverged from one another. Laurasiatheria and Euarchontoglires undoubtedly form a single larger northern group, but whether Afrotheria or Xenarthra first branched off, or whether the tree forked in two with Afrotheria and Xenarthra sharing their own unique common ancestor, remains debated (see Figure 9.3).

Additionally, the precise branching pattern within parts of Laurasiatheria remains opaque, as do a few other branch points, not least the question of where exactly the bats fit in.* But the most prominent dispute concerns when the major divisions occurred. What is at stake is not only when the four groups originated but when exactly the last common ancestor of all extant placental mammals lived.†

---

* Bat history has always proved tricky; in addition to their complicated genetics, they appear rather fully formed in the fossil record, with no known intermediate forms leading up to them.
† Equally lively debates exist regarding the exact phylogeny of marsupials, especially concerning relationships between Australasian and American tribes.

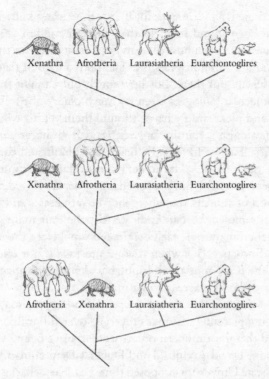

Figure 9.3: Uncertainty remains as to the exact relationships between the four main clades of placental mammals.

A happy birthdate

Molecular studies have a habit of estimating that the lineages they identify originated at much earlier dates than any palaeontologist ever suggests. The rate of change of DNA sequences – the molecular clock that Linus Pauling and Emile Zuckerkandl developed – appears to be much slower than the fossils would otherwise suggest.*

---

* Inferring historical genealogies has always been an exercise in probability, but the current maths is mind-bending. Molecular phylogeny is today a discipline dominated by higher mathematical modelling and complex statistics, with the assumptions of the models used profoundly affecting the outcomes. I once stood at the back of

But there is an issue with fossils. Say the earliest known fossil bat had been found in rocks that were 52 million years old. This doesn't tell you how old bats *are*; it only confirms that bats are *at least* 52 million years old. The odds are good that older bat fossils are out there. But how much older might they be? The molecular biologists often say much older – the folks with spades and picks simply haven't found them yet. To which the palaeontologists typically answer, 'Sure, we haven't looked *everywhere*, but we have had a really good look, and bats can't be *that* much older …' It can get tense. And this isn't just about bats or mammals. Similar exchanges occur regarding the evolution of animals, the emergence of vertebrates and so on.

All the molecular data indicate that the four main groups of placental mammals each originated way before the demise of the dinosaurs. But when it comes to fossils, not one from before the KT boundary indisputably identifies an ancestor of a living order of placental mammals.

Following a DNA-based proposal from 1999, suggesting that living placental mammals diverged over 100 million years ago and that most modern orders were in place before the KT boundary, David Archibald and Douglas Deutschman, at San Diego State University, proposed three possible scenarios. They called them the 'short-fuse', the 'long-fuse', and the 'explosive' models (see Figure 9.4). The first two scenarios accept that ancestors of current placental mammal groups lived deep in the Mesozoic, whereas the latter says that just one lineage leading to modern mammals survived the mass extinction and only then, well, explosively diversified in a dinosaur-free world.

The fossil record seems to support the explosive model. And a 2013 study that amalgamated morphological and molecular data also supported it, suggesting that all placental mammal groups were born post-dinosaurs. Numerous molecular geneticists have, though, attacked this work for giving unreasonable importance to the morphological data

---

a seminar while the speaker discussed the age of placental mammals and watched this family tree expand and contract like an accordion across sequential slides that each employed a different methodology.

and consequently requiring that mammalian genomes evolved after the dinosaur extinction at rates typically only associated with viruses – 60 times faster than the standard animal rate.

Opposing the explosive model are numerous studies of genes that indicate an origin for placental lineages well within the Mesozoic. The short-fuse model then says that the ancestors of today's major placental mammal lineages all emerged very soon after the first placental mammal lived. The most extreme study favouring this scenario suggested that the disappearance of dinosaurs had only a tiny effect on mammalian diversity.

Conversely, the long-fuse model asserts that the major branches of the mammalian family tree were established before the dinosaur extinction, but they persisted at only low levels of diversity before flourishing after the meteor struck. With this model, the lineages might have split genetically but without morphologically diverging, so they were not recognisable as ancestors of their living descendants.

Hence, the short-fuse model suggests that many Mesozoic mammal fossils have been missed; the long-fuse model argues that many fewer fossils of much rarer animals have eluded discovery, or current fossils have been misidentified. Sometimes, people talk about a 'Garden of Eden' hypothesis, whereby all these mammalian orders would have evolved in a place that's never contributed to the fossil record.

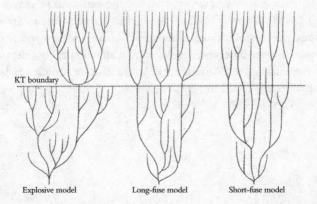

Figure 9.4: The nature and timing of the placental mammal radiation remains debated.

This three-way debate probably has plenty of fuel in the tank. Recent fossil surveys further support the reality of a post-dinosaur explosion, but few researchers deny that the fuse of that blast was lit in the Mesozoic. On balance, the long-fuse model seems to hold the best cards. Plus, some new readings of DNA have estimated a somewhat more recent – but still Mesozoic – origin of placental mammals.

The one thing most researchers agree on is a familiar scientific mantra: more data are needed. Plus, molecular biologists and palaeontologists need to keep finding new ways of collaborating and fitting their contributions together – and the latter probably do still need to find even more fossils.

The mammalian family tree that we might confidently draw is, of course, a cartoonish sketch of a 310-million-year reality, but it is an amazing achievement. This tree's form results from evolution's inherent branching and mammals' intrinsic biology having collided with myriad external influences – two catastrophic mass extinctions (plus a good many smaller ones), the evolution of the dinosaurs, the demise of the dinosaurs, the arrival of flowering plants and, later, of grass, the changing weather and the tectonic thrusting of continents.

When Charles Darwin folded his hypothetical phylogeny into *On the Origin of Species*, he left the root of it blank. Some things, he believed, were apparently too deeply buried in history to ever be known. I think he'd appreciate how deeply we have traced mammalian history, and how ceaselessly his successors try to fill the empty spaces.

CHAPTER TEN

It's Getting Hot in Here,
Put Your Coat On

A herd of thousands of wildebeest stretches back from the water's edge. The lush green-bordered river cuts through boundless darker yellow-brown-green plains. The heads of the wildebeest front row are bowed drinking, and, as snouts rhythmically twitch, eyes unblinkingly scan the water. There are crocodiles in this river. And they are hungry.

Wildebeest must drink, though. Plus, as they perpetually circle Kenya and Tanzania searching for rain-fed fresh grass on an annual 1,600-km (1,000-mile) migration, they must occasionally cross a river. How different their lives are to those of the crocodiles, who simply wait for nomadic ungulates to arrive.

It is the largest crocodile that leads the hunt. His tactic is to approach by stealth, then surge from the water. So far, he has made some unsuccessful lunges, each ambush having failed because of the wildebeest's lightning reflexes.

Perched in a tree on the opposite bank, I hold my breath and feel my hands tighten on my binoculars. I fumble to focus on what's either a floating log or the huge croc again, drifting towards the bank.

Only kidding. If my grip tightens on anything, it's pizza. But Cristina and I are sitting bolt upright on the sofa, rooting for our mammalian cousins as their murderous cold-blooded pursuers stalk them. As David Attenborough narrates, we greet every crocodilian failure with huge relief.

But it isn't a log. And SNAP! The croc gets what he wants – the wildebeest's left back leg is clamped in his jaws. The crocodile pulls his victim onto its back, then quickly into and under the water. From all directions, younger crocodiles skim across the river to aid in the kill. Chunk by chunk between a dozen jaws, the wildebeest is torn apart. 'Crocs can't chew, so they have to spin together to tear pieces off the carcass,' says Sir David, and we watch a giant tail rise from the river, then thrash back down.

Finally, a shot of the first assailant. Triumphantly, he tosses his head back to swallow a wildebeest limb whole. He doesn't even come close to chewing. And as the hoof disappears, Attenborough delivers his closing zinger: 'He won't feed again until the wildebeest return next year.'

When I discuss this sequence with people the following week, it's this last comment that we talk about in disbelief – '*Was that right?*' '*One meal a year?*' It *is* correct. The largest of crocodiles can go a year without eating. A bellyful of wildebeest is enough to feed their cold-blooded metabolism for 12 months, leaving just enough to fuel the next annual hunt.

By contrast, if a common shrew goes more than five hours without eating, it dies. Shrews eat two to three times their own body weight every day. 'Shrew life,' says the *Encyclopaedia Britannica*, 'consists largely of a frenetic search for food.' This is the price of being warm-blooded.

Of course, there are other differences between a Nile crocodile and a shrew. Most pointedly, the former is 5m (16ft) long and weighs 700kg (1,540lb), while the latter measures about 7cm (less than 3in) and tips the scales at around 10g (0.35oz). And when it comes to body temperature, size matters.

Nevertheless, more equitable comparisons are still striking: Mammalia's top predators, tigers, need a good meal every two or three weeks, not once a year. Whereas a leopard gecko, somewhat akin to the shrew in size, can go weeks without food.

The general rule of thumb is that, compared to a cold-blooded vertebrate of similar dimensions, a warm-blooded mammal or bird consumes up to 20 times more calories.

Watching ever more dynamic and high-octane animals emerge on the fossilised path from early amniotes to the first mammals is exciting, but their lifestyles were increasingly expensive – 20 times more calories is no minor hike in the energy budget. Today, even to sit still costs mammals significant energy. Determining what drove our ancestors to assume such an expensive physiology has, therefore, long vexed biologists. The first question is, what exactly could have made this greater expenditure worth it?

We're still not sure. Part of the problem is that warm-bloodedness lies at the heart of mammalian life. In influencing just about every aspect of our physiology and behaviour, it gives theorists a lot to choose from in terms of explaining its importance.

What is warm-bloodedness?

In academia, 'warm-blooded' and 'cold-blooded' are now sternly frowned-upon colloquialisms, terms that lost their scientific credibility in Victorian times. The main problem is that so-called 'warm-blooded' animals can have cold blood – while hibernating, for example – and 'cold-blooded' animals frequently have warm blood. Iguanas, by dashing to and fro between sun and shade, adding and removing heat-absorbing pigment from their skin, and regulating blood flow to their surfaces, have excellent control of their warm body temperatures (at least while the sun is out).

Today's preferred descriptors are 'endotherm' and 'ectotherm'. Endotherms produce heat internally through their own metabolism, while ectotherms rely on external heat sources to warm them. Typically, when people talk about endotherms, they mean birds and mammals.* There are constantly comparisons to be made between these two lineages, and the parallels – developed independently – are remarkable.

---

* Numerous other organisms generate heat to warm themselves. But it is typically a transient or localised phenomenon. Sharks, tuna and swordfishes all generate internal heat, often targeted to specific organs such as the brain, eyes or muscles. Certain insects internally generate some heat too, as do several plants.

Another important concept is 'homeothermy', which describes the maintenance of a constant body temperature. Bird and mammals are therefore endothermic homeotherms.

Temperature is fundamental to biology because all organisms are ultimately just localised masses of chemicals reacting according to the enzymes present, enzymes being proteins that catalyse chemical reactions. And neither those chemicals nor the enzymes escape the basic laws of physics and chemistry, which state that temperature determines how quickly things happen and that hotter = faster.

Being warmer, therefore, increases an animal's ability to do stuff. Moving, dodging predators, hunting prey, digesting, thinking, growing, reproducing and so on – all of these processes proceed faster in a warm body.[*] And obviously, doing such things faster than your competitors is typically advantageous. That is evident in the rapidity with which wildebeest leap back from the river to avoid crocodiles.

Homeothermy, by contrast, in maintaining a constant temperature creates a stable environment for a body's countless chemical reactions. Different enzymes and biochemical processes are not equally sensitive to temperature – an increase of just a few degrees can speed one reaction more than others. Because life depends on these reactions being integrated with one another, a constant temperature facilitates the coordination of interacting chemical processes.

Most mammalian body temperatures are held constant at a value between 35°C and 38°C (95–100.4°F). And being able to maintain a constant warm internal temperature is key to mammals inhabiting a broad span of climates.

Endothermy works by the animal generating excess heat, and this heat flowing from the animal to the outside world. This process requires that the animal has (1) a means of heat generation, and (2) control of the resultant thermal energy. In birds and mammals, heat is produced by metabolism in the

---

[*] Albeit only up to a point; most proteins start malfunctioning at about 45°C. (113°F)

visceral organs – mainly the intestines, liver, kidney, lungs and heart – and this heat's efflux, or outflow, is primarily regulated at the body surface by insulating coats of fur or feather.

The idea that heat *loss* is fundamental to maintaining a warm temperature is counterintuitive. Feathers and fur, one typically thinks, act to stop heat escaping. More accurately, though, these outer coverings *slow* the escape of heat. By thermal energy being continuously generated and its flux actively regulated, a dynamic process is created that is responsive to an ever-changing world. Fur and feathers don't just slow heat escape by a constant amount; they can be manipulated to allow heat to pass at different rates, so as to regulate body temperature.

Insulation is so important to the slowing down of heat loss that here another great conundrum in explaining the evolution of endothermy is thrown up. If increased heat production, with its inescapably high energy costs, occurred before the evolution of hair in mammals, then that heat would have been quickly lost to the wider world, inflating even further the costs while deflating the benefits.

What's more, it seems impossible for ectotherms to have evolved hair because it would have stopped vital external heat from getting in to warm them. Indeed, experiments done by Raymond Cowles at the University of California in Los Angeles in the 1950s showed that dressing monitor lizards in tailor-made fur coats disrupted their ability to thermoregulate.

The movement of hair adjusts the rate of heat loss by determining the thickness of the layer of trapped air around the animal. If a mammal is cold, it stands its fur on end to create a deeper layer that acts as a greater barrier to heat escape. When humans get 'goose bumps' in the cold, it's a rather sad attempt to expand the fur coats we lost maybe a million years ago.

Large mammals living in very cold places, such as reindeer, arctic foxes and polar bears (which are black below their white coats) have long, dense fur that allows them to create large volumes of warmed air around themselves. Their hair is typically 3–7cm (1–3in) long, so this isn't an option for smaller mammals, which would trip over it. Small mammals in cold places tend to look for sheltered warmer microclimates, often below the snow or in burrows. Huddling is important for these animals. In northern Canada and Alaska, five to ten taiga voles will share a nest, and go out to forage at different times so the nest is always kept warm by their shared body heat. And, when temperatures get too cold, there's always the option of temporarily abandoning endothermy by hibernating through the worst of times.

In addition to adjustable insulation, mammals also regulate how much blood passes near to their surfaces. To promote heat loss, superficial blood vessels are expanded; to prevent heat loss, the vessels are constricted. This and hair movement are low-cost strategies for dealing with minor temperature fluctuations.*

When body temperature shifts are more extreme, however, emergency measures are taken. A fall in temperature first boosts the metabolic rate to increase heat production. If this isn't enough, shivering can generate heat from rapid muscle contractions. Plus, mammals possess a unique type of fat tissue called brown adipose tissue, the only function of which is to generate heat. It's especially abundant in newborn, small and hibernating mammals. And crucially, its heat production can be stimulated on demand.

Conversely, if the body is in danger of overheating, many – but not all – mammals sweat. We humans certainly do. Sweating requires particular skin glands, those mammalian specialities, that release watery secretions onto the skin, which cool the body as they evaporate. Dogs can only sweat through their paws. Instead, like a lot of other mammals, they

---

* Mammals, such as otters and beavers, that have assumed semi-aquatic lives have oily fur to keep water out of their trapped air layers, whereas whales and seals have convergently hit upon blubber as a form of insulation. It's an excellent barrier to heat, but you can't move blubber up and down. Instead, redirected blood flow is important for heat loss – warm blood being rerouted via vessels nearer to the surface or through the blubber-free fins.

pant so that water evaporates from their tongues to help cool the body.

If these are the standard-issue mechanisms by which mammals cool themselves, those that live in the hottest of environments have had to concoct even more elaborate schemes. Because endothermy requires heat loss, it's tougher to be warm-blooded than cold-blooded in hot places – if heat can't escape, hyperthermia ensues. Cooler burrows and shade, in general, are important; as is an ability to operate in the cool of the night. But certain desert mammals – including that emblem of endless sandscapes, the camel – actually relax their temperature regulation. They've found ways to tolerate their bodies warming by as much as 5°C (9°F) over the course of a day.

Not that camels don't have a number of other adaptations to thermoregulate. They have only thin hair on their bellies, so if they find a cool surface (most likely in the early morning) they can lie on it to lose heat. If the air is warmer than their body temperature, they huddle with other camels to stay *cool*. And they have a remarkable tolerance for dehydration. The urine of dehydrated camels is apparently 'dark brown and syrupy'. And when camels reach water, they can drink 30 per cent of their body weight in 10 minutes. Drinking so much water that quickly would kill most mammals, but camel physiology is specially adapted to deal with the shock.

If a constantly warm body temperature is elemental to mammals living across a huge variety of habitats, it's not without attendant hard work.

To generate heat, visceral organs have essentially become less chemically efficient. The twin fuels of metabolism are glucose from food and oxygen from breath. Their inherent energy is transferred into the chemical bonds of a molecule called ATP, which can be spent by the body at a later time. But in birds and mammals, the membranes of the mitochondria, the sausage-shaped powerhouses of our cells, have become leaky, and ions move across them to generate heat without contributing to ATP production. This is all rather complicated, but the key messages are: mitochondria generate chemical energy *and* heat.

And the more heat or ATP is made, the more oxygen and calories from food are burnt.

That brown adipose tissue I mentioned, through which mammals store energy and release heat on demand, takes things a step further. The mitochondria there contain an 'uncoupling' protein that means they barely make any ATP as they throw out heat.

The rate at which an organism consumes oxygen at baseline is called its 'basal metabolic rate'. And it's the BMR that has been raised way up high in mammals, so that heat spills out of the viscera and warms the body. Mammalian BMRs are about 10 times higher than in comparably sized reptiles.

Why the BMR increased so dramatically is the primary question that's addressed in theories of endothermy. Finding an advantage for increasing *basal* metabolic rates – that is, for burning more energy while essentially doing nothing – is difficult. If there are clear advantages to mammals and birds being much more active than ectotherms, why not just develop the ability to be highly active when necessary, while still resting with an economical BMR?

We'll deal with various theories, but whatever the answer, vast swathes of mammals' biology have been sculpted by their high-energy lifestyles: many unique or highly specialised mammalian traits are adaptations to the need to acquire and process enormous amounts of food and oxygen.

In addition to the lung-boosting diaphragm, specialist biting and masticating teeth, and the nose's respiratory turbinates encountered in the last chapter, we can add specialised hearts, kidneys, blood and guts.

Mammals (and birds) have four-chambered hearts, unlike the three-chambered ones of other reptiles and amphibians. The extra chamber came from splitting in two the bottom chamber that pumps blood out. Once two output chambers existed, they could form the basis of separate circulatory systems – one to the lungs and one to the rest of the body. The latter could supply the body with blood at a pressure higher than the lungs could tolerate, and so deliver oxygen to where it was needed faster and more effectively.

High blood pressure also powers filtration by the kidneys, which in mammals (and birds) reach great sophistication. Hugely increased metabolic rates generate a lot of waste that the kidneys must remove. In mammals, urea is dissolved in water, but the kidneys then – by incredibly elegant means – reabsorb water to produce concentrated urine.

Also, the oxygen-carrying cells of the blood are special in mammals. To pack as much haemoglobin into red blood cells as possible, these cells have lost their nuclei. Hence, these cells – produced at a rate of about two million per second in adult humans – live for three to four months without any genomic DNA. In mammals (and birds), red blood cells have also become very small, to speed the transfer of oxygen in and out of them.

Finally, the mammalian gastrointestinal system – into which hi-tech jaws and teeth deliver well-chewed food – is another organ system that's been tweaked in a number of ways to ramp up efficiency.

All these systems point to only one conclusion: mammalian life burns fast.

When did endothermy evolve?

In the previous chapter, we saw how the fossil record of mammalian ancestors depicts a series of progressively more active animals. To determine when mammalian lineage first contained bona fide endotherms, students of endothermy need to decide which transformations along this history robustly indicate that shifts in metabolic status had occurred.

Starting with our oldest ancestors, pelycosaurs, there is the evidence that their spinal columns straightened to allow them to breathe while running. But little else indicates a spike in activity levels.

Things get much more interesting with the next protomammal grade, the therapsids. These animals evolved the bony secondary palate separating their mouths and noses to allow them to breathe while eating. And their rib arrangements suggest a diaphragm had evolved. Then there are the air-warming, water-saving respiratory turbinates in the nose.

When Willem Hillenius claimed that these structures would make an excellent proxy for the evolution of endothermy, nobody argued with the logic, but they did question his suggestion that some ridges on a therapsid's nasal cavity indicated that the animal had once held extensive turbinates. However, in 2011, a German research group examined the noses of *Lystrosaurus*, that barrel-chested, pig-like therapsid that was the most common animal on earth after the end-Permian mass extinction. Analysing a fossil with 'neutron topography', they found *Lystrosaurus* had had respiratory turbinates made of cartilage. These therapsids lived about 250 million years ago, suggesting that endothermy may have evolved 30–40 million years before true mammals arrived.

This early date got further support at the start of 2017, from an ingenious new approach to this problem. Adam Huttenlocker and Colleen Farmer, working at the University of Utah, picked up on the fact that avian and mammalian red blood cells are much smaller than those of ectotherms, and that these cells pass through correspondingly small capillaries. They then reasoned that if they looked at capillaries in fossilised bones, they might see a decrease in the vessels' diameters that would signal the evolution of high-energy metabolism. And, indeed, they saw such a reduction. It was observed, this time, in a cynodont – the next grade closer to true mammals – but an early cynodont. The age of these fossils? About 250 million years.

Together these studies provide pretty compelling evidence for therapsids and early cynodonts having taken significant strides toward endothermy. The next logical question is whether therapsids had hair?

Alas, with hair and skin we're back to structures that only fossilise under exceptional circumstances. The earliest unequivocal evidence of hair comes from the remarkable north-eastern China fossil beds but dates only to about 165 million years ago. That is well past the birth of mammals, and most investigators believe fur must pre-date this.

Certain investigators have claimed that pits apparent on the snouts of some therapsid fossils would have contained whiskers, indicative of at least those sorts of hairs existing. This controversial claim may be supported by recent evidence that these snouts contain space for nerves – potentially carrying sensory information – to run back from the pits.

We do know, however, that at least one therapsid alive 255 million years ago was hairless, as it left behind a very clear rocky impression of its skin. So detailed was this skin imprint, described in 1967, that this fossil might also say something enlightening about where hair came from. The impression seems to indicate that the animal was covered in small glands, tiny little structures that looked as if they secreted something onto the skin.

Hair is obviously one of Mammalia's most definitional features. The usefulness of a full coat of hair and its transformative effect on the ability of mammals to thermoregulate are indisputable. What we know is that hair is made of keratin proteins, which became vital in the days of early amniotes waterproofing their skin. Keratins have served amniotes well: besides hair, they've been employed in the construction of scales, claws, nails, hooves, horns, beaks (bird, turtle and platypus) and feathers.

But a fur coat is another biological trait where you scratch your head and ask: what good could have come from 10 per cent of this? Consequently, theories of how hair first evolved seek to identify a structure from which it may have developed that didn't originally function to retain heat.

The two most promising ideas were proposed more than 100 years apart. The first dates from the late nineteenth century but was developed most thoroughly in 1972 by Paul Maderson, of Brooklyn College, New York. Maderson suggested that hairs were originally sensory appendages. The central idea is that protohairs were fine protruding structures that, attached to sensory nerves, were highly sensitive to being deflected. Among today's amphibians and reptiles, there are hairless creatures that have functionally similar sensory bristles or spikes. Perhaps, for therapsids, such structures were useful additions to the animal's sensory arsenal. Then, Maderson suggested, these initially sparse structures multiplied according to chance mutations until they became dense enough to provide some functional insulation.

A newer thesis, more firmly rooted in the development of hairs, emerged in 2008 from Kurt Stenn and colleagues, at the University of Pennsylvania, and in 2009 from Danielle Dhouailly, of the University Joseph Fourier in Grenoble. Stenn called it the 'sebogenic hypothesis', but more informally it's referred to as the 'wick hypothesis'.

A hair is no simple structure; what sticks out of the surface is only one component of a pretty intricate little construction. The protruding hair is a collection of dead cells pushed up from a follicle, which contains a collection of renewable stem cells for making hair. A small muscle is attached to the shaft to move the hair up and down; and leaking into every hair follicle is a sebaceous gland, whose function is to secrete an oily lubricant onto each hair. We encountered these glands when considering the origins of mammary glands.

Stenn wondered whether in reptiles, the challenge of waterproofing the skin had been met by hard scales while in the mammalian ancestors who'd retained glandular skin, a water-repellent oily secretion had become a crucial part of skin biology. Stenn suggested that deeper, more profuse sebaceous glands were favoured, and that hair evolved as components of these, acting as wicks to raise the oily protectant onto the skin.

Danielle Dhouailly's case for hairs arising from sebaceous glands came from a broader examination of how vertebrate skin develops. She writes that the entire skin of a mammal is by default programmed to make hair – it only doesn't if hair production is actively suppressed, as happens on the palms of hands and the cornea, for example. This suggested that hair evolved from a structure that covered the entire bodies of mammalian ancestors, which sebaceous glands likely did.

Further, while no hair follicle lacks a sebaceous gland, sebaceous glands do exist without hairs in them – on the lips, eyes and genitals, for instance. In fact, the signal that tells a region of skin to make a hair is required at a high dose to make both a hair and a gland, whereas a low dose of it induces only the gland. Together, these observations suggest that hair evolved as an auxiliary component of a pre-existing gland. First, it wicked the escape of a secretion; later, it assumed an insulating function.

That living monotremes, marsupials and placental mammals share similar fur dates hair production to well before the last common ancestor of living mammals, and it's almost certain that the tiny first mammaliaforms and mammals couldn't have survived without insulation. Some researchers think that the furry monotremes might provide clues to the endothermic biology of early mammals. Their body temperatures tend to fluctuate a little more than most mammals', and to average out at a slightly lower level – somewhere in the low 30s °C (high 80s °F). Echidnas also hibernate and enter periods of so-called torpor – long-term and short-term bouts of hypothermia, respectively. They might also use their behaviour to aid thermoregulation. Again, monotremes help us imagine transitional states – this time – between ectothermy and endothermy.

Why did endothermy evolve?

Biologists have appreciated for centuries that warm-bloodedness is a defining characteristic of mammalian and avian life, but contemporary discussions of why endothermy evolved rarely reach back further than the mid- to late 1970s, when three influential theories were proposed. These models have been celebrated, attacked and debated ever since, with just one major new angle added in 2000. Considering how different they all are, it might seem strange that none has yet been consigned to the rubbish bin of bad ideas, but we'll get to that.

Brian McNab, at the University of Florida, proposed in early 1978 that 'miniaturisation' was pivotal to the evolution of mammalian endothermy. Here, we're back to the crocodile and the shrew. Because small animals have a lot of surface area compared to volume, they gain and lose heat very quickly. For small ectotherms, this means their bodies will warm quickly in the sun, but cool quickly too. If a small ectothermic animal increased its internal heat generation, that heat would rapidly be lost, making it hard to believe endothermy's first steps could have been taken in small bodies.

Conversely, the larger an animal gets, the smaller its surface becomes relative to its bulk; therefore, there's a comparatively smaller area for heat exchange. (Bulkier animals also have more muscle and tissue in general, which impedes heat loss.) As a consequence, large ectotherms – such as adult Nile crocodiles – can maintain fairly constant and high body temperatures by default. They're called *inertial homeotherms*, this warmth and thermal stability existing simply due to their size.* And the diminishment of relative surface area with increasing size is why warm-blooded elephants, hippos and rhinos, for example, don't need fur to impede escaping heat. (At least they don't today – living in an ice age, woolly mammoths obviously needed 'wool'.)

McNab proposed that in the run-up to mammals evolving, large therapsids became inertial homeotherms. And that as therapsids, cynodonts, and mammals became progressively smaller, natural selection favoured innovations that sustained the high body temperatures their larger therapsid ancestors had become accustomed to. This process culminated in fur and full endothermy.

Why animals on this lineage were getting smaller then becomes a parallel debate – the majority opinion is that this was to make use of ecological opportunities open to more diminutive creatures, perhaps due to the emergence of dinosaurs. More problematically, while some therapsids were large and early mammals were definitely small, the fossil record does not show a straightforward decrease in size: therapsids came in all shapes and sizes, and even cynodonts were a variable bunch.

A second idea was that endothermy permitted mammals to explore a greater range of 'thermal niches'. Relying on external

---

* Large dinosaurs were likely inertial endotherms, although their thermoregulatory physiology is a can of worms that will remain firmly closed here.

heat sources – primarily sunshine – to stoke physiological fires limits the times and places in which ectotherms can be active. Might endothermy have created independence from external temperatures? Was its evolution favoured because it allowed protomammals to be active at any time of the day or night, or to invade colder climates? Certainly, most evidence points to early mammals having been nocturnal, potentially to avoid the dinosaurs which dominated daytime proceedings.

The third and most influential idea of the 1970s, however, was published in 1979 by Albert Bennett, at University of California, Irvine, and John Ruben, of Oregon State University. Bennett and Ruben proposed that ideas based on natural selection directly favouring animals with controllable high body temperatures looked at the problem the wrong way around.

Endothermy is costly, full stop. But theories positing that heat generation and warm bodies were selected before fur and other regulatory mechanisms existed to somewhat reduce costs require that a hugely expensive process had benefits that outweighed the masses of resources it would have demanded. For example, in 2000, to show just how much energy would have been needed to raise an ectotherm's body temperature, Bennett substantially increased the metabolic rates of monitor lizards by force-feeding them a large meal, but saw the animals' body temperatures barely budged. Trying to raise body temperature first would have been like putting on the central heating while leaving the windows open.

Bennett and Ruben proposed, instead, that endothermy was a by-product of nature favouring animals with abilities to maintain higher levels of activity; specifically, those animals whose activity levels had been increased by evolving larger 'aerobic capacities'.

There's not much to separate the short-term sprinting abilities of reptiles and mammals. David Attenborough narrates another brilliant sequence in which a freshly hatched iguana heroically dashes away from an ambush by racer snakes – neither the snakes nor the young iguana are slouches over short distances. But this sort of frantic activity is powered *an*aerobically. In times of need, the body rapidly burns up available chemical energy without consuming oxygen. Conversely, more sustained periods of physical activity have to be powered by real-time oxygen consumption. Bennett and Ruben suggested that natural selection had acted on animals' abilities to maintain oxygen-powered high-energy movement over a passage of minutes or hours. They showed that in an hour an iguana could maybe cover half a kilometre, but a similarly sized mammal could cover four.

Selection for aerobic capacity made good sense, but Bennett and Ruben still had to explain why this would necessarily lead to endothermy. To do so, they sought a relationship between aerobic capacity and BMR. There wasn't a huge amount of data to choose from; measuring aerobic capacity is an involved process where animals in gas masks are placed on treadmills. Plotting available data, however, revealed that in ectotherms and endotherms alike, maximal aerobic power was always approximately 10 times higher than the BMR. The two traits seemed fundamentally linked. If so, selection for greater aerobic capacity might simply have inescapably inflated BMR and hence led to endothermy.

The immediate question is, of course, why are these two processes linked? Bennett and Ruben suggested that the traits selected to improve aerobic performance – such as greater abilities to absorb and transport oxygen, more mitochondria and their improved efficiency – would have had the off-target effect of inflating overall metabolism.

There's a lot to like about this theory – there are tangible benefits of higher activity levels for eluding predators or catching prey, and investigations continue to seek further evidence in support of it – but it hasn't been universally accepted. The chief issue is how certain it is that maximal aerobic consumption is always tenfold higher than the BMR. The basic correlation has not been thoroughly refuted, but there are some notable exceptions to the rule, suggesting that aerobic capacity can be increased independently of baseline metabolism. For example, pronghorns – the North American equivalents of Old World antelopes – appear to be able to consume oxygen at 70 times their basal rates, while alligators running on treadmills raise their consumption fortyfold – and who knows what they'd achieve swimming?

Critically, during exercise, the extra oxygen is not being consumed by the visceral organs that mediate baseline metabolism, but by muscle. Perhaps, there are fundamental links between the two tissues – shared genetic control of mitochondrial number and function, say, or common mechanisms of oxygen transport and absorption – but many investigators have asked why the two systems couldn't have evolved independently. Couldn't evolution have created animals capable of great aerobic ability when they needed it, but who rested without expending masses of energy? As Colleen Farmer – then at University of California, Irvine – put it in 2000, 'there is no mechanism that explains why endothermy would be essential to sustain vigorous exercise.'

This remark came in the introduction to a paper entitled '*Parental care: the key to understanding endothermy and other convergent features in birds and mammals*', where Farmer argued that rather than endothermy having been the unifying feature that drove the emergence of the many similarities between birds and mammals, parental care was the shared starting point for the emergence of their overlapping biology. Most pertinently, parental care, Farmer said, had driven the parallel evolution of warm blood.

Farmer began by discussing the advantages of incubating offspring at elevated and constant temperatures. Farmer describes how moderate temperature shifts – easily tolerated by adult animals – can kill embryos, and that less severe temperature shifts can still cause serious developmental defects. Additionally, the fact that heat speeds things along would have meant that warmed embryos developed quicker, meaning less time spent in an egg – a popular meal for scavenging carnivores – and reproductive maturity would have arrived more quickly.

What makes Farmer's argument compelling is the examples she compiled of ectothermic animals that today utilise different heat sources for reproductive purposes. A common theme among both vertebrates and invertebrates is the building of nests to retain moisture and heat. But also, from caterpillars that pupate sooner if they huddle together, to lizards that sunbathe and rush to their eggs to transfer the heat, to maternal pythons who wrap themselves around their egg clutches and shiver to release heat, there are numerous examples of warmth being funnelled to developing young.

Among mammals, sloths and tenrecs – African animals that resemble a cross between a mouse and a hedgehog – are not the best of endotherms: their average body temperatures are low-ish and fluctuate more than most. But during pregnancy such slovenly physiology isn't tolerated – they raise both their body temperatures and their regulatory game. Similar shifts are seen in egg-laying echidnas and hummingbirds.

Farmer suggests that thyroid hormones, which regulate metabolic rate, may have been central to switching on a primitive, long-term but reversible form of endothermy. The warmth would have been expensive but of great benefit to the young. This theory got a serious boost in 2016, when a living lizard species was found to become transiently endothermic while it was reproducing.

Moving to the post-hatching stage, parental investment is stretched even further. Farmer notes that every day, great tits make nearly a thousand trips to their nests with food for their offspring, and that, to feed their offspring, lactating mammals increase their usual daily energy budget by four to ten times. As we saw in Chapter 5, the benefits of accelerating development are clear – the less time spent being a vulnerable youth, the better. Farmer proposes that sustaining the constant physical activity required to achieve development-quickening, parental provisioning would have driven the evolution of greater aerobic capacity.

In considering postnatal care, Farmer's thesis overlaps with a second parental care hypothesis of endothermy published in 2000. Pawel Koteja, at Jagiellonian University in Kraków, independently proposed that it was this need to gather enough energy for the substantial task of looking after hatchlings that drove the evolution of endothermy.

Like Bennett and Ruben's aerobic capacity model, Koteja's focus was on natural selection favouring increased levels of activity. But Koteja argued for a different timescale. To forage for food for their young, parents had to sustain elevated movement levels not over minutes or hours, but over hours or days. This required extra eating and, crucially, to benefit from extra food the body needed to digest it and assimilate the energy quicker. Hence, raised metabolic rates would have been selected in the intestines, liver and kidneys – exactly the organs that today maintain a high BMR in mammals and birds.

At the heart of this idea, the animal entered into a positive feedback loop – better energy assimilation required increased metabolism, which facilitated better energy assimilation; increased overall activity improved food capture, which allowed increased activity.

This sort of positive feedback may play into any theory of endothermy's evolution, but Koteja focused on parental care, writing, 'Evolutionary ecologists agree that decreasing juvenile mortality and accelerating growth to maturity are among the most effective ways of increasing fitness.'

In suggesting that parental care provoked the evolution of endothermy, Koteja proposed that a behavioural change might have laid the cornerstone for subsequent physiological and anatomical changes, which is a fascinating idea. It does, though, present the toughest problem of his and Farmer's theories. Throughout this book, I've bemoaned the fact that soft tissues don't fossilise; well, good luck finding a fossil trace of a behavioural change …

It's also worth acknowledging another implication of these theories. Today, over 200 million years after these critical transformations happened, I take great satisfaction in shopping and cooking for my family. I helped pay the heating bills when Cristina was pregnant, and I took her blankets when she was on bed rest. (I know, it was the least I could do!) I have thought, while making soup, about how mother and father birds share parental feeding responsibilities. Farmer's and Koteja's theories, applied to birds, affect both sexes equally. But recall that fatherly care is rare in mammals and that it was almost certainly not the done thing among early mammals. Neither Farmer nor Koteja is explicit about this, but their work suggests that the selective forces that drove the mammalian lineage towards endothermy would have acted on only half of the population. If genes for endothermy were selected for when they were inside mothers, maternal biology would have been key to the evolution of one of Mammalia's most defining characteristics.[*]

One for all, all for one ...

The various proposals for endothermy's origins have evoked some pretty fierce debate over the decades, and academics slugging it out over their ideas is a wonderful spectacle. But recent thinking about warm-bloodedness has become much more magnanimous, almost reaching a point where people are unwilling to entirely reject any of the major ideas. Indeed, when Koteja addressed an international meeting regarding endothermy in 2004, rather than presenting his own ideas, he chose to consider how theory itself had evolved since the 1970s. 'Perhaps,' he said, '[...] the proposed models focus on different aspects of the process rather than provide progressively better explanations.' Then, in a fascinating analysis, he described how the theories, in having grown to incorporate different aspects of biology, reflected shifts in evolutionary thinking.

The earliest theories, he began, proposed that homeothermy alone was selected for, to provide biochemical stability, and so they focused solely on the *intrinsic biology* of the organism. Next, the idea that endothermy allowed the animal to be active in different thermal niches looked at the interaction between the organism and its *physical environment*. Following this, the aerobic capacity and miniaturisation models – in, respectively, focusing on prey–predator interactions and ecological competition – considered organisms' interactions with their

---

[*] This said, sons inheriting greater aerobic capacities or abilities to forage further would have put these traits to use in seeking out further mating opportunities. And, then, if those males bred more successfully, their daughters might have inherited the genes necessary for better parenting ...

living environment. Finally, his and Farmer's work said it was insufficient to consider only adult animals; evolutionary physiology must consider the *entire life history*. 'Individuals do not come to the world as "plug and play" or "turn-key" devices,' he stated. 'Natural selection acts along the entire path, but the selection mesh may be finest early in life.' The different theories were not necessarily opposed, they might just be considering distinct aspects of the same multifaceted phenomenon.

In 2006, Tom Kemp expanded on this theme, developing ideas he'd first explored in his 1982 book *Mammal-like Reptiles and the Origin of Mammals*. Any theory that attempted to pin the emergence of endothermy on a single cause, Kemp said, was doomed to fail. Endothermy was too complex and its effects on an animal's biology too pervasive for its evolution to be attributed to one factor. Kemp saw a morass of physiological and biochemical processes that contributed to both BMR and high aerobic capacity, meaning that endothermy and peak activity levels were, indeed, fundamentally linked. And a change to any of those processes would concurrently shape an animal's behaviour, parenting capabilities and locomotor possibilities. 'Virtually everything in the biology and life of a mammal,' Kemp states, 'is either contributory to, or affected directly or indirectly by, the endothermic temperature strategy.' Therefore, there was never a physiological change that influenced only parenting or hunting capabilities, say. There was never a single advantage of endothermy that occurred first to be followed only later by all the other consequences of endothermy that are apparent today. Instead, protomammals and mammals evolved on multiple fronts in parallel; an irreducible system affecting the whole animal had incrementally moved from ectothermy to endothermy.

Regarding the sorts of events that might have led to endothermy, Kemp proposes a hypothetical mutation that slightly increased mitochondrial numbers. If the cardiovascular system supplied enough oxygen to power more mitochondria, this mutation would have fractionally boosted aerobic capacity and therefore increased possible activity levels somewhat. But more mitochondria would have slightly raised body temperature too, and allowed the animal to operate a little later into the night, plus it would also have increased parental stamina a bit. However, a second mutation further increasing the number of mitochondria would have had no further effects if oxygen levels became a limiting factor; say, for example, the lungs weren't big enough, or the heart wasn't strong enough. Then, a mutation increasing oxygen delivery would have to had come next. A complex system is only as strong as its weakest part allows. A mutation favouring greater parental attentiveness would only make a difference if it occurred in a body capable of going the extra mile for the kids … We like to talk about chicken-and-egg scenarios, but Kemp is saying the whole concept of one wondrously complex creation preceding another intricately linked one should be abandoned altogether. Any two or more interdependent biological traits help make one another.

Finally, in 2016, Barry Lovegrove, of the South African University of KwaZulu-Natal, claimed that single-cause explanations of endothermy had created a 'conceptual stasis'. Then he embarked on a 300-million-year history of endo-thermy, from which he concluded that, in both birds and mammals, the evolution of this complex trait was marked by three recognisable 'pulses'.

Phase One lasted from 275 million to 220 million years ago and, for the mammalian lineage, corresponded to the evolution of therapsids. The array of changes that marked the evolution of these ancestors – their upright gaits, expanded jaws, specialist teeth, diaphragms, and, in particular, those respiratory turbinates – was taken as evidence that therapsids were highly active animals operating at elevated temperatures. In the absence of any compelling indications that these animals had fur, behavioural measures to aid temperature control might also have been important. Lovegrove suggests this phase was driven primarily by greater parental care, homeothermy and greater aerobic capacity – answers to challenges established when tetrapods had crawled up onto dry land.

Phase Two, from 220 to 140 million years ago, was crunch time. This era was when body sizes dropped, brains grew dramatically, and mammals crucially gained fur. Also, it's when a nocturnal existence was critical to mammalian survival. Because size is so important to temperature control, Lovegrove views mammals' miniaturisation as pivotal to this pulse, but adds that parental care and aerobic capacity would still have featured in the selection landscape. This upgrade to endothermic biology and its inherent value may well have contributed to the surge in mammalian diversity in the Jurassic period.

Phase Three is focused on differences *between* mammals. Certainly monotremes maintain lower body temperatures than other mammals, but marsupials are on average a couple of degrees cooler than placental mammals too. Larger differences between mammalian orders began to accrue after the dinosaurs departed. Following this cataclysmic event, certain lineages evolved even stronger endothermic capabilities. In particular, mammals that galloped and those that invaded colder climates developed higher metabolic rates. For example, Lovegrove has found that the faster a mammal runs, the higher its BMR tends to be. This finding invokes Bennett and Ruben's model, but Phase Three came long after the origins of endothermy.

The wildebeest we began with are among the quickest of mammals and have body temperatures at the higher end of the spectrum. The 'wait and pounce' mode of hunting employed by the crocodiles who prey on them likely resembles the way early aquatic and semi-aquatic tetrapods caught food. It's a system seen more rarely in today's higher-energy world.

We first met Barry Lovegrove in Chapter One. He made the recent and strongest case that testicles were externalised because the abdomen became too warm for sperm production. Always, in that chapter, I thought of the scrotum as a solution to a problem: things got too warm or too bumpy, and so the scrotum evolved as a sanctuary for male gametes. Problem solved, end of story. In Lovegrove's long view, though, the evolution of the scrotum is a liberation. As body temperatures rose, problems with spermatogenesis temporarily halted further rises; so the scrotum's evolution didn't simply fix a problem, it freed animals to keep getting warmer.

Lovegrove also believes that hibernation and briefer bouts of hypothermic torpor, when body temperature is allowed to decrease dramatically, have always been a part of mammalian biology, although others have viewed these temporary abandonments of endothermy as recently evolved adaptations to cold weather. And Lovegrove makes a remarkable claim based on this view: the mammals that survived the meteor blast that exterminated the dinosaurs might simply have slept through it!

He's not the first person to suggest that the mammals that survived the initial impact and the immediate infrared blast of heat were in their burrows. But Lovegrove also thinks these energy-saving tactics might have been useful in navigating the chaotic post-apocalyptic world.

When Albert Bennett reviewed evidence bearing on his and Ruben's aerobic capacity theory in 1991, he commented on how hard it is even to understand the selective pressures acting on living species, and therefore it's 'probably foolhardy to attempt it for unknown organisms now extinct that lived in poorly understood environments'.

It would be nice to have more theories in biology whose elegance came from their simplicity. But, unlike the unchanging nature of matter itself, which means E always equals mc^2, the way matter is arranged in living systems changes. That's the very basis of the evolutionary project. Lovegrove's 2016 account is more of an attempt at a historical record than a scientific theory. Kemp too appeals to the contingency of history in his forward walk of many small steps. He likens the quest to pin endothermy on a single cause to failed attempts to find a single cause for vertebrates moving to the land. But just as tetrapods gradually transitioned from aquatic habitats to drier ones through a spectrum of intermediate habitats, mammalian ancestors probably moved from low- to high-energy lifestyles through a series of medium-energy ones. Single-cause theories appeal to our aesthetic sense – they contain that 'Ta-dah!' moment – but human tastes don't get to determine history.

CHAPTER ELEVEN

Scents and Sensibility

One of the greatest cons of my childhood, zoologically speaking, was that people (no one in particular, just people generally) led me to believe that most animals went about their lives as humans do, waking for the day and sleeping at night. I thought only a small band of mysterious animals, led by owls and bats, lived by night. And I believed, too, that these creatures possessed quasi-magical abilities to see in the dark – how else, after all, could you function without light?

So entrenched was this view, it was remarkably resilient to new conflicting information. As I grew up, I came to know that nearly all my compatriot mammals – foxes, badgers, hedgehogs, rats, mice, even rabbits and hares to some extent – were nocturnal. But not once did I question the view that nocturnality was a minority sport practised only by eccentrics.

This belief was then further reinforced by the way people spoke about dinosaurs forcing early mammals into the night. It always seems implicit that mammals were obliged to make do with a second-rate ecological niche. Consequently, one tends to assume that once the giant reptiles shuffled off their mortal coils, mammals would have immediately resumed daytime living. In the movie, you imagine the dinosaurs dead, and the gentle smile of a relaxed mammal feeling the sun warm its face for the first time.

Such were these convictions that recently – aged 39 – while inspecting the wide-ranging collection of mammal specimens at Oxford's wonderful Museum of Natural History, I was shocked. One after another, the labels above the mammals described them as nocturnal. So many of them said this I wondered if there could have been a mistake.

Of course the labels were accurate, and a survey from 2014 quantified exactly the extent of mammals' affinity for the night. Considering 3,510 species, this exercise showed that only 20 per cent of mammals are – like humans – diurnal, 8.5 per cent are active regardless of the time of day, 2.5 per cent come out during the twilight of dawn and dusk, and so nearly 70 per cent of all mammals are nocturnal.

The only defence of my childhood notion lies in birds nearly all being diurnal, apart from owls, and the majority of reptiles – a rare sight in my corner of rural England – also being out by day, solar-powering their metabolisms. Plus, it is accepted that diurnality was the ancestral habit of tetrapods and amniotes.

Really, though, today's preponderance of nocturnal mammals shouldn't have been a surprise; just as all mammals have mammary glands because they descend from the same lactating ancestor, this last common ancestor was nocturnal too. And because of the persistence of diurnal dinosaurs, mammals spent a considerable chunk of their history unable, or unlikely, to live by day. One hundred and thirty million years is a long time to be specialised for functioning by the moon rather than by sunlight:* changing sleep–wake cycles would have taken significant re-adaptation.

Rather than mammals entering the day *en masse* after dinosaurs exited, it happened sporadically in different lineages. Another survey that plotted the activity patterns of 700 mammals on the mammalian phylogeny estimated that various mammals have ceased to be nocturnal around 16 separate times.

Diurnal living has spread most widely among the ungulates and across our tribe, the primates. Otherwise, over 1,000 species of bat remain night-fliers; and despite a few lineages becoming diurnal, most of the 2,000-plus species of rodent are also still nocturnal. Insectivores tend to be active 24/7 – as we know, shrews need to eat constantly. More surprisingly, tigers, lions and a clear majority of other carnivores are also

---

* Long enough to lose a system seen in other animals that repairs UV-damaged DNA.

creatures of the night, as is the black rhinoceros. The ant-eating numbat is the only truly diurnal marsupial. Old habits, it seems, die hard.

If endothermy was and is mammals' means of coping with the chill of darkness and their initial entryway into the sanctuary of night, adopting this lifestyle also demanded that mammals managed with little or no light, a fact that profoundly affected the way they sensed the world.

Senses

Another childhood memory. Every Sunday – from about four years old to twelve, when football took over that day – my dad took me walking in the local countryside. We'd cross farmland to reach mature woodlands, then loop back through more fields. Always, we watched for animals. Mainly, we'd see rabbits and birds, some species of which were a treat, but the most exciting spot was always deer. Foxes and badgers, of course, were asleep. When we were in deer territory, dad would hiss 'Shush!' We'd then stand dead still and scan the forest for one of these animals. The goal was always to *see* deer – only catching sight of one mattered. A crackle of sounds unmistakably made by a large, fleet-footed animal wasn't a sensory interaction. It was a near miss.

To a deer though, a human voice or the sound of heavy footfall is enough – with that, they vanish. A deer's nervous system isn't wired to value the *sight* of a potential predator. I'd walk on, acutely aware of how impossible it was to move without making a sound; of how hopeless a goal silence was.

Humans tend to think of themselves as having five senses: vision, hearing, smell, taste and touch. All of these considerably pre-date mammals – some sense organs are vertebrate inventions; others are even more archaic. What concerns us here is how mammalian biology was affected by the refinement and specialisation of a handful of these ancient channels of information.

Vision, hearing and smell are *distance* senses: reflected photons, travelling waves of air pressure and volatile chemicals move from their sources to our eyes, ears and noses to inform us about what exists beyond the boundaries of our bodies. In contrast, stimuli physically contacting our bodies trigger taste and touch.

Taste is a relatively limited sense: receptors on the tongue sample ingested material for just a handful of signals (most of what we humans perceive as taste comes from the smell of food). For a long time, sensory biologists thought we cared only how sweet, salty, bitter or sour our food was, but we now recognise a fifth quality. Tongue receptors also report on a food's glutamate content, evoking a taste of 'umami' – Japanese for 'pleasant savoury flavour'.

Touch, in its simplest sense, reports on the shape, texture and movement of objects contacting the body. An accumulation of touch receptors in certain body parts, such as lips and fingertips, make these excellent instruments for exploring objects' properties. The possibilities of exploratory touch are nowhere more impressive than in the reading of Braille.

But to adhere to this notion of five senses, we must subsume to touch the sense of how hot or cold an object is. And what about pain? And itch? These two senses blur the line between internal and external stimuli – the bite of a predator hurts and the bloodsucking of a mosquito itches because there are nerve fibres that report on whether the body has been injured or agitated. Pain and itch are concerned with the *effects* of predators and parasites. Opening the door to senses concerned with the state of the body, we must consider that we are equipped with sensors for blood pressure, sugar level and oxygenation, and systems for knowing how inflated the lungs are, how extended the gut is and how hard the heart beats. Nerves run from all our voluntary muscles to indicate how flexed they are, so that the nervous system can know where the body's limbs are in space. The mechanisms that ears use to detect sound waves evolved from a system for reporting the position and movement of the head, an inherent sense of balance. On and on it goes.

Honing all of these internal senses doubtlessly accompanied the evolution of mammals' relentlessly high-energy lifestyles.

Here, though, we will deal with only the externally oriented senses – the means by which an animal surveys its wider circumstances.

The point of any sensory system is that it provides information on which an organism does well to act. A wildebeest leaping back from the water's edge when it sees a crocodile is not so conceptually different from a bacterium changing the genes it expresses to digest nutrients it senses in its environment. Organisms can only respond – behaviourally or physiologically – to what they can sense, and this is determined by the detector systems they possess. These systems can vary according to *what* triggers them, and in how *sensitive* they are. With the evolution of sensory organs (and the brains they feed information to), we typically see burgeoning abilities to extract ever more information from the torrents of photons, sound waves and chemicals that the organism encounters. Or, indeed, they evolve to sense other forms of tenable information – recall the platypus's bill sweeping a riverbed, responsive to electricity emanating from the muscular contractions of its prey. What's important is that the information reaching the nervous system says something useful for the survival of the organism. The platypus finds its dinner; the deer flees from the sound or scent of a potential predator.

The reason a child, or even an adult human, finds the idea of a nocturnal existence so alien and unreasonable lies in the richness of our visual streams. Standing in a forest, we see in brilliant detail the tapestry of trunks, branches and twigs; we perceive an immense palette of browns, oranges and greens; when animals appear we watch them intently. To stand in a forest and close our eyes – or to stand there at night – what are we left with? A disorientation of bird calls, sounds of flapping wings, rustles of leaves ... and I have no idea what a forest smells like. Almost always, my visual stream feels like the sensory world to which all other inputs are rooted, a frame of reference that other sensory streams only supplement, the way a film's sound engineer feels secondary to its cinematographer.

But this is a hopelessly anthropocentric viewpoint.

Sight

In mammals, sight is a diminished sense. When proto- and early mammals moved into the night, numerous components of the ancestral tetrapod visual system were purged. For example, it's popular among fish, amphibians and reptiles to have a 'third eye', the pineal eye, a spot of neural tissue on the top of the brain typically covered by skin but not bone, that directly senses ambient light levels to entrain brain activity and hormonal levels to the time of day. Mammals lost this structure.*

Also, the shape of most mammalian eyes and the dimensions of their various parts speak of a nocturnal history. They have large corneas to permit large pupils capable of flooding the retina with light. The typical mammalian eyeball is shaped so that light needs to cross only a short distance to reach the retina, meaning it doesn't spread widely over many light receptors. Mammal eyes also have a huge preponderance of rods. Rods are one of two types of light receptors in a vertebrate retina. They are the highly sensitive ones, specialised for low-resolution vision in dim light conditions. A 2016 study showed that mammals had devised a special developmental pathway to create extra rods.

Most shockingly, however, to members of a species that delights in colour, mammals have historically neglected this aspect of perception. Colour vision is mediated by the other light-responsive cells of the retina: the cones. Cones need greater stimulation than rods, and they come in different varieties, each maximally sensitive to a different wavelength of light. By comparing the activation of different cones, the brain creates a sense of colour. Humans have three distinct cone types – one activated by longwave light that we see as red, one activated by the medium wavelengths perceived as green, and a shortwave receptor that evokes blue.

Given that different objects – foodstuffs, predators and potential mates – reflect and absorb different wavelengths of

---

* Birds and snakes have also lost these third eyes.

light, this system allows the animal to distinguish objects by their perceived colour. But here's the rub: most non-mammalian tetrapods have not three but *four* types of cone. And most non-human mammals have only two.

An examination of vertebrate genes for cone pigments – the receptor proteins that determine to which wavelength a cone responds – reveals that early tetrapods inherited four such receptors from fish and that most reptiles and birds have retained these. Looking at the genes in living mammals, however, indicates that early mammals probably only had three – a medium-wavelength receptor was probably lost early in protomammalian history.[*] And, fascinatingly, the monotreme lineage then lost one shortwave receptor – evident as the dilapidated remains of that gene in the platypus genome – while therian mammals lost the other shortwave receptor.

Mammal eyes also lack the coloured oil droplets that facilitate sensing different hues in other vertebrates, confirming the relegation of colour vision. Most mammals, therefore, have rod-dominated retinas supplemented by two cones. The retention of these short- and long-wavelength cones does suggest that this reduced system remained useful, but most mammals have only a rudimentary sense of colour, akin to red–green colour blindness in people. This state of affairs may account for how lacking in colour mammals are. Unlike birds and reptiles with their four cone pigments, and amphibians with three, all of which have over the aeons evolved plenty of colourful representatives, the palate employed for making mammal fur is decidedly autumnal.[†]

---

[*] In fact, a 2014 study examining fossil eye shapes suggests certain pelycosaurs and therapsids may have been nocturnal long before dinosaurs came along.

[†] It's still strange that there are no green mammals. Some sloths appear foliage-coloured, but it turns out their greenness is due to algae that live on them.

The fact that humans have three types of cone and pretty decent eyes is a result of primates' reinvasion of the diurnal niche. Along this trajectory, the gene for the longwave cone pigment duplicated, and the resultant pair of receptors diverged in their wavelength sensitivity. Within the primate family, there are many experiments in re-establishing intricate colour vision with primate visual systems having reinvented numerous functional components that mammals had abandoned for more than 100 million years.[*]

Hearing

It's time to pick up again the mammalian jaw joint and those two small, exorcised bones left hanging two chapters ago. Earlier, when thinking about placentas, I remarked how with hearts or hands, say, that object's function was writ large in its visible structure, but that with brains or placentas, the relationship between structure and function was only apparent at the microscopic level of cells. The ear – viewed with a pretty low-grade microscope, or even a good magnifying glass – is in the same category as the hand and heart. The ear's structure is exquisitely revealing of its function. There is a visible elegance rather than microscopic mystery (although it has that too). One can almost imagine its moving bony parts as the creations of a brilliant craftsman, each one shaped and filed to make a beautifully artisanal sound-detecting device, especially when it comes to the middle ear of mammals.

An ear is typically divided in three, and the outer, middle and inner ears of mammals are each unique.

The outer ear is the part on the outside of the head – which people colloquially call the ear – plus the external auditory canal, a tube that carries sound waves towards the

---

[*] Whether this state of affairs is directly responsible for the bright blue scrotums of vervet monkeys is moot, but the red, white and blue faces of mandrills certainly seem like something only a colour-sensing animal would appreciate.

eardrum at the canal's end. Mammals are the only animals that have sound-funnelling structures, 'pinnae', on the outsides of their heads. That's how much they value hearing. Or at least how much therian mammals do – platypuses and echidnas, like non-mammals, lack these accessories. Humans' external ears are pretty weird structures and not especially functional. The muscles, for instance, which allow other mammals to point their ears towards where sounds are coming from, no longer work in people.* Plus, most terrestrial mammals have considerably larger pinnae than humans do. Hares' ears are ridiculously large, and those of many bats, foxes and, indeed, deer are often not far behind. These structures aid hearing in the same way as an old-school hearing trumpet does, by herding sound waves towards the auditory canal and the eardrum.†

The eardrum marks the start of the middle ear. But let's first consider the inner ear: this is where the business of converting changes in air pressure into a signal to be sent to the brain occurs. The cells that do this are termed 'hair cells', and they lie in sheets with their hair-like protrusions extending into the fluid-filled tubes that constitute the inner ear.

In the different tubes of the inner ear, the fluid moves in response to discrete events. One part is concerned with balance and movement, and it is the animal's movements that shift the fluid there. This tracking of the body's motion was probably the original function of the ear, evolving prior to vertebrates heading landwards.‡ Hair cells in the hearing part of the inner ear are stimulated by travelling waves of vibrations – sounds – moving the fluid. When the movements

---

* Besides, that is, those few people who retain enough functionality to wiggle their ears and often, in my experience, like to demonstrate this ability.

† Elephants have giant ears too, but while these animals do have excellent hearing, their flappy ears are enlarged mainly to help cool their large frames.

‡ Fish also have lines of hair cells running down their bodies, which detect external water currents.

of this fluid bend the hair cells' hairs, they spark an electrical signal in the cell, which is the currency of the nervous system. Hair-cell structure is apparent only in the microscopic realm, but hair cells are further evidence that structure defines function, and the triggering of neural signals by the bending of their hairs is a reminder that sound is always about motion.

The inner ears of mammals are distinguished from non-mammalian ones by the extended and coiled tube in which sound-evoked vibrations are detected. In mammals, this structure is called the 'cochlea', after the Latin for snail. Guinea-pig cochleas make four full turns; in platypuses, they make just a half turn; and human ones move through three-and-a-half spirals. The size and geometry of the cochlea are, in part, elemental to mammals being able to hear a very wide range of pitches, and, in particular, to the mammalian speciality of being able to hear considerably higher-pitched sounds than other animals.

But also fundamental to this ability is the nature of the mammalian middle ear, a structure that is one of Mammalia's most definitional traits. To understand the middle ear, we must again consider vertebrates moving onto land. When aquatic vertebrates first dedicated hair cells to detecting externally derived signals, they needed very little by way of an amplification system. Sound waves are stronger and travel faster in water, so they easily propagate through to the fluid of the inner ear. A sea is apparently a very noisy place if your ears are adapted for picking up waterborne sound, but ours are not. An air-to-fluid transition is a wholly different matter.

For terrestrial animals to hear, they had to invent a new system for detecting and *amplifying* much weaker airborne vibrations. In fact, what amounted to hearing in early land-dwelling vertebrates was the detection of ground vibrations conducting up through their limbs and lower jaws to provoke their ears. Chewing, when Mammalia's ancestors developed it, would have been disturbingly loud.

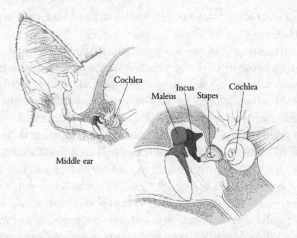

Figure 11.1 The mammalian ear with its three middle ear ossicles.

The task of amplifying airborne sound waves – so that they were strong enough to send fluid waves running through the inner ear – was solved by the evolution of eardrums coupled to vibrating bones in the middle ear – the so-called ossicles. Ultimately, terrestrial vertebrates would have large eardrums that vibrated bones of certain shapes, so that those bones then tapped hard enough with their other ends on the inner ear. The ears of frogs, reptiles and birds achieve this with just one ossicle. Whereas mammals have their definitional three middle-ear bones – these ossicles lying in a beautifully organised chain that is excellent at amplifying vibrations, and that is especially useful for hearing high-frequency sounds.

First in the chain, attached to the mammalian eardrum, is the hammer, or 'malleus' in Latin. The hammer strikes the anvil – 'incus' in Latin – and the anvil moves the stirrup – the 'stapes' – which stimulates the inner ear. It's this chain that evokes, for me, the idea of an eclectic craftsman deeply versed in sound engineering designing this ear. But, of course, the challenge with biology is never to decipher the mind of a creator but to uncover the incremental workings of blind evolution. And the central conundrum of the mammalian middle ear has always been: how did the extra two bones get there?

Remarkably, the most important insight regarding this question came in 1837, the year Darwin was just sketching his famous '*I think* ...' sketch in Notebook B. Then, Geoffroy Saint-Hilaire – whom we met as a misguided agitator in the debates over platypuses – was a much more influential figure than young Darwin. Geoffroy asserted that all animal bodies adhered to a single archetypal body plan, which meant all body parts of any given animal had to have equivalents in other animals – no body part could be unique to a particular creature. The data acquired to support this notion would later aid enquiries into homology and evolution – but Geoffroy sought equivalence between body parts in a non-evolutionary framework based on his idea of a single master plan. The three bones of the mammalian middle ear presented a particular challenge to this world view, and a good number of pre-eminent anatomists took up the challenge of seeking the bones equivalent to the mammalian ossicles in other animals.

The crucial insight came from German anatomist Karl Reichert after he'd taken his dissecting needles to pig embryos. To decipher where the ossicles came from, Reichert traced their developmental trajectory, and by doing so, he correctly inferred that the hammer and anvil were equivalent to jaw bones in non-mammalian vertebrates. More precisely, Reichert saw that the hammer was akin to the articular bone of the jaw, and the anvil to the jaw's quadrate; the two bones that had formed the abandoned jaw joint of pre-mammalian ancestors, as discussed in Chapter Nine. Reichert also said the stapes was equivalent to the single ossicle in frog and reptile middle ears and, further afield, comparable to a structural support in fish skulls.

When evolutionists later picked up the story of the middle ear, they initially thought that early tetrapods had first evolved an ear-drummed middle ear, using a superfluous cranial support bone incorporated into that space – that former cranial support having become a single ossicle, the stapes – which transmitted vibrations to the inner ear. They also believed that mammals had later inserted a further two bones to improve the dynamic range of this structure. This made intuitive sense, but it was one heck of an engineering problem – how could a functional single-boned ear reasonably insert another two elements between the eardrum and the existing ossicle?

The short answer is that it didn't. As investigators looked more closely at living tetrapods' ears, they noticed telling differences between those of frogs, those of reptiles and birds, and those of mammals. Notably, mammalian eardrums seemed to be in a different position. Also, as palaeontologists dug deeper into protomammal fossils, there was no evidence of a middle-ear form intermediate between reptiles' and mammals'. More shockingly, there was also no evidence of mammalian ancestors having had single ossicle middle ears.

The consensus view now is that tetrapods evolved ears based on tympanic eardrums and vibrating middle-ear bones on at least three separate occasions: once in frogs, once in early reptiles and once in mammals, mammalian ears owing their particular form to the evolution of the jaw joint. Remarkably, this theory says mammalian ancestors went some 100 million years without a freely moving ossicle-based mechanism for hearing airborne sound, then early mammals inserted three bones into a sophisticated middle ear in one fell swoop.

This theory of mammalian ears was cogently argued for the first time in 1975 by Edgar Allin, at the University of Wisconsin, who surveyed a range of protomammal and mammal fossils to suggest that, in therapsids, an eardrum formed over the three future ossicles while they were all still part of the jaw. Just because they were part of the jaw didn't mean, however, that they couldn't contribute to hearing. The articular and quadrate – the future hammer and anvil – may still have been parts of the jaw joint, but they were, by this point, small bones in contact with the stapes, which pressed against the inner ear.

These bones and their auditory potential were then liberated by the formation of the new mammalian jaw joint. Once the teeth-holding dentary docked on the skull directly, the articular and quadrate were free to work their new jobs full-time. Gradually, the chain of three ossicles migrated away from the jaw to form a more independent ear, which could be further insulated from the noisy disruption of chewing.

All this means that not only are the mammalian middle-ear bones former jaw bones, but the contact between the hammer and anvil in each of our ears is homologous to the jaw joint of ancestors and of living reptiles.

The long process of forming the mammalian middle ear began with selective forces honing the masticatory finesse of mammalian jaws and decreasing the size of two once-prominent jaw bones. But in the latter stages of this transformation, it's much harder to allocate primacy to selection for jaw function or the attainment of better hearing. Biologists have argued in favour of both; either way, mammal hearing got much better.

For insect-eating nocturnal animals, this must have been very useful. The advantages of good hearing at night are fairly obvious: the sounds of predators' approaches and of insects would have been crisper and louder. Plus, hearing has natural advantages over vision. It is impossible, for instance, to see around a corner or through a tree trunk, but while photons travel in straight lines and are easily blocked, sound waves pass around (and even, to some extent, through) solid objects. Arctic foxes hunt by listening for mice invisible to them below the snow, then they leap to nosedive into the white based on what their ears tell them.

These arguments, however, principally relate to more sensitive hearing *per se*, and not specifically to why mammals profited from hearing more acutely at higher frequencies.

One important aspect of hearing is that it's often used to receive vocal signals from members of one's own species. And mammals – as noted earlier – are often sociable and always heavily invested in taking care of their offspring. The tiny mammals who evolved high-frequency hearing were almost certainly capable of only very squeaky vocalisation and, so, might these animals' ears have been sculpted by improving capacities to hear each other's calls?

As useful as communication is, it can also draw unwanted attention. A youngster calling to its mother, for instance, might attract a predator's attention – but only if the predator can hear the cries. The evolution of high-frequency hearing likely provided early mammals with a significant opportunity: the ability to make and hear sounds pitched higher than those detectable by larger – say, dinosaurian – animals' ears would have provided early mammals with their own private communication channel. They could converse in high-pitched sounds, their neighbours oblivious to their exchanges. Indeed, today, what we humans perceive as the high-pitched squeak of a mouse is actually at the low end of the murine repertoire; most mouse vocalisations are too high for us, or any reptilian or avian predators, to hear.

However, while this conversational quirk was, and is, doubtless useful for mammals, another factor probably propelled the evolution of high-frequency mammalian hearing. For hearing to be really useful, an animal needs to know from where a given noise is coming. And just as having two eyes facilitates depth perception significantly, having two ears hugely increases the three-dimensionality of sound. By the brain's auditory processing centres comparing differences in the timing and intensity of inputs to two ears, the origins of sound waves can be computed.

For an elephant, or indeed a person, the head can provide a big enough distance between the two ears to make a sound arrive at the ear furthest from the sound's source with a significant delay. But for a shrew-sized creature, that delay is tiny. The alternative is to use differences in intensity. For long sound waves, the signal at either ear is barely different, but the shorter the sound wave, the higher the frequency and pitch, the more different it will be at the two ears. Consequently, if small early mammals were able to hear higher frequencies, they would have had a greater chance to compute from where that sound came. Supporting this there is a striking correlation between mammal head size and hearing – the smaller a mammal's head is, the higher the pitch it can hear.

Another thing that helps mammals to localise sounds is the outer ear – the pinna – which is especially good at indicating whether the sound is coming from your back or front, from above or below. All those years ago, a deer in a wood, big pointed ears adjusting, would have known exactly where my father and I were coming from.

It would be wrong to end without acknowledging the most impressive innovation to have arisen from mammalian hearing: sonar.

Nineteenth-century investigations into bats' nocturnal navigational skills began somewhat brutally with experiments showing that blinded bats retained these skills, while deafened ones were left helpless. But in the 1940s, Donald Griffin carried out the key experiments demonstrating that bats 'echolocated'. At Harvard University, Griffin revealed that bats emitted short ultrasonic pulses of sound, then listened to the echoes of their calls returning to them. By this means, bats navigate their environments and catch flying insects, increasing the rate at which they emit sounds up to 100 times a second as they home in on their prey. These animals, at least, really do have a quasi-magical ability to see in the dark.

Sonar is also a good strategy for life underwater. Convergently with bats, toothed whales – such as sperm whales and orcas – and dolphins (which are strictly toothed whales too) also emit high-frequency clicks, then pay attention to what bounces back, their ears having readapted to the properties of external water rather than air.

Whales are also the loudest of animals. The low-frequency clicks sperm whales use for communicating are now known to be even louder than those of the blue whale. These calls reach 230 decibels and carry through the ocean for thousands of miles.

Another quirk of toothed whales is that they've abandoned a sense that mammals spent a good deal of time elaborating. These animals have no ability to smell.

Smell

In Alaska, a polar bear has been observed walking 65 kilometres (40 miles) directly to seals that he had presumably smelled. Bears – not too far from bloodhounds on the phylogenetic tree – are the doyens of mammalian smell. And mammals generally have excellent senses of smell.

Humans are decent enough with our noses, but our expanded visual abilities seem to have come at some cost to this sense. We do, however, still respond strongly to certain odours. Whether they shock, delight, repulse or sexually arouse us, a central feature of scent perception – scientifically, 'olfaction' – is that many smells are hard-wired to evoke an innate emotional response. People, though, rarely explore new surroundings by actively sniffing them, and we don't typically greet each other with a thorough nasal inspection. In contrast, many mammals use smell much more to locate food, parents, offspring and sexual partners, to avoid predators, and to mark their territory and intimidate others.

Being able to sample chemicals that waft about you is a very ancient sensory capacity. Mammals have not added significantly to the basic mode of doing so; they make receptors that can be activated by passing molecules carrying potentially valuable information, and these receptors are, as in all vertebrates, concentrated in the nose. The story here is again one of mammalian augmentation rather than outright invention.

One indicator of olfaction's high standing among mammalian senses is the large amount of tissue dedicated to it. Olfactory sensory cells spread to form a thin sheet in the nasal cavity and, using a time-honoured mechanism for increasing the amount of a slim layer of tissue that fits into a cramped space, the sheet in mammals is crumpled. Its area expands by placing it on elaborately folded bones – the olfactory turbinates. Just as respiratory turbinates create a large surface area for the water content of air to be controlled while breathing, further nasal turbinates provide a greater surface area for attaching tissue that detects odours. While paleontologists haven't found sensory turbinates in the fossilised ancestors of mammals, from pelycosaurs onwards fine bony ridges are seen where olfactory cells probably once lay, indicating that mammalian ancestors probably had olfactory turbinates and a keen sense of smell from early on.

Most mammals also dedicate a significant amount of their brains to processing inputs from their noses. Because expanding a brain region increases its computing power, the value an animal assigns to a particular sensory channel can be inferred from the neural real estate it allots to it. In humans, for example, visual processing occupies large swathes of the brain, whereas in mammals generally, olfactory brain regions are especially large. However, the starkest sign that mammals value their noses highly is the sheer number of olfactory receptors they make.

Humans' technicolour vision is the result of signals from just three different colour sensors in our eyes – the countless hues we perceive arise from comparing the signals from that trio of receptors. To uncover the molecular logic of smell likewise required determining how many different olfactory receptors there were. From the late 1980s, a number of people working in Richard Axel's lab at Columbia University tried to do this. The person who succeeded was Linda Buck, who in a nocturnal burst of creativity, in 1991, isolated the genes for rat olfactory receptors. Examining the huge mass of genetic data she'd accumulated, Buck and Axel didn't find three olfactory receptors, or 10, or 100. The data indicated that rats had about 1,000 of these receptors. The revelation that smell operates by mammals having many, many receptors won Buck and Axel a Nobel Prize.

Subsequent work has shown this estimate was about 200 short of the real number in rats. It has also taught us that some receptors seem to be activated by just one chemical, while others are generalists, sensitive to various odours.

Other vertebrates have considerably fewer olfactory receptors than mammals. Totals vary, but lizards and fish have about 100, birds and turtles around 200, while alligators reach 400. When exactly the expansion happened isn't entirely clear. Within placental mammals, having just over 1,000 receptor genes seems to be about the norm – this number has been found in a variety of species, including horses, cows, rabbits, dogs and various rodents. Remarkably, in these species, olfactory receptor genes account for a full 1 per cent of the entire genome.

Looking beyond placental mammals to inform a possible phylogeny of olfaction, the marsupial opossum also has about 1,000 receptors, while platypuses have about 350. This likely means that a major receptor expansion occurred after the monotremes diverged but before the split between marsupials and placentals. Although, it's conceivable that the semi-aquatic, electrosensing platypus lineage may have secondarily tuned down olfaction. The fact that olfactory receptors come and go astonishingly frequently confuses the matter – as mammals diversify, new receptor genes come, and old genes go at surprisingly fast rates. Receptor repertoires of living mammals are hugely variable, and genomes are littered with the remains of former receptor genes.

Exceptions to the 'roughly 1,000' rule among placental mammals include those toothed whales, primates and elephants. It seems toothed whales no longer found smell useful living their aquatic lives – today, these animals lack olfactory nerves, the brain structures that analyse smell are gone, and nearly all their olfactory receptor genes have degenerated. In primates, the diminishment of olfaction is subtler. Humans are considerably less endowed with receptors than rodents, but we still have 400 or so functional ones. In elephants, though, the opposite is true. These animals have about 2,000 functional olfactory receptors (and 2,000 broken ones). Elephants have long been known to 'smell' water from miles away, but discovering this genetic abundance was a surprise. The researchers who found it offered, however, some examples of elephantine olfactory prowess: male elephants, when they're in aggressive phases, release signals from scent glands behind their eyes, and African elephants can recognise up to 30 different family members by smell. They can also sniff the difference between two sets of tribespeople, telling apart those who like to spear elephants and those who peacefully farm around them.

Smell is a very social sense, as well as being, like sound, a good night-time sense. It's also useful to animals whose noses are close to the ground – meaning small mammals, like rodents, rather than elephants. Just as the properties of sound waves and photons mould hearing and vision, the functions of olfaction are shaped by the features of odorous chemicals. Seeing and hearing are concerned with fleeting, real-time signals – signals that die as soon as they are born – whereas smell is a much slower sense – odorants diffusing from their sources over much longer timeframes.

Just as a sense of smell is an archaic trait, so too is odour-signalling between animals. But mammals, owing to their glandular skin, are highly adept at it. In addition to milk and sweat, scents are an essential secretion of mammalian skin. Scent glands are not only in elephant eyes; they exist on the flanks of insectivores, the anuses of most carnivores, and in beavers' castor sacs (again near the anus), the products of the latter being used in human perfumes. Male ring-tailed lemurs compete to impress females by having 'stink fights'. Each male prepares by smearing his tail with the secretions of two glands – a watery one from glands on his wrists and a 'brown toothpaste-like substance' from his shoulder glands. The two males then wave or flick their tails at each other until one backs down. The people who've studied this display, however, report that to them the smell is imperceptible. Like the ultrasonic chatter of rodents, smell opens the possibility of private exchanges between members of the same species.

Good human olfaction is a mammalian inheritance; its latter-day shortcomings might be a price for the switch to better vision, or perhaps high up in trees smell was of less use to primates. There is, though, plenty of evidence that the scent of a potential reproductive partner consciously or unconsciously remains important to people. This may be related to the compatibility of immune systems in making strong offspring. In fact, one theory for why humans retain the curious patches of body hair they do, in their armpits and groins, suggests that they help waft important olfactory cues from these body parts.

Touch

Although smell is important to a mouse, when its snout is busily twitching about its world, the animal is not just sniffing – its whiskers, too, are generating a plethora of sensory information. As one commentary, put it: 'Eyes may be "the windows to the soul" in humans, but whiskers provide a better path to the inner lives of rodents.' To end this survey of mammalian senses, let me briefly consider touch, and specifically touch mediated by those most distinctive of mammalian appendages – hairs.

Touch is another primordial sense – organisms of all stripes sense pressure bearing down on their bodies – but mammal skin is laden with many types of specialised touch receptors, and the most sensitive of all their detectors of mechanical displacements are those that enwrap the ends of hairs. Like the protrusions of the ear's hair cells, hairs act as useful levers for sensing movement. And the whiskers of rodents and other mammals represent the zenith of exploiting the potential of such a structure for exploratory touch. A scurrying mouse navigates with its whiskers, recognises objects with them and uses them in social exchanges. Its whiskers are much more densely supplied with nerves than other hairs, and the pattern of whiskers is mapped perfectly on the mouse's brain in a region called the 'barrel cortex', a barrel of neurons for each one. It hardly needs stating that such a system is very useful in the dark of night, but it is even more valuable in the absolute darkness of the subterranean burrows in which many rodents reside.

While highly specialised 'tactile hairs' are limited to the faces of most mammals, they are occasionally more widely distributed, such as on the wings of bats, or across the whole body of the otherwise furless manatee. With this in mind, we return to Paul Maderson's argument from 1972 that hair originated not as insulation but as a sensory device. Maderson's thesis was that to get from hairlessness to a coat of fur useful for being endothermic – to cross the sort of gap St George Mivart thought was unbridgeable – perhaps hairs evolved first to aid touch and then multiplied to cover the body. In this case, smaller protrusions would have been useful sensory levers, and this advantage could have driven hair's early development; there would have been a clear usefulness in having even a sparse number of them. This model suggests, again, that a shift in function occurred during a complex trait's evolution. Insulation, and hence endothermy, may have been born of the usefulness of sensing the world more fully.

Other worlds

Aged four, I stood in forests desperate to catch sight of whatever had just made the sound we'd heard. I focused all my concentration on my eyes scanning the view. Now, I stand in forests trying to impress on my children how lucky we'd be to see something as majestic as a deer. It's my turn to hiss, '*Shush!*' And still, of course, a sight of a deer is what matters most. Now, though, I appreciate my environment a little better. Dense woodland, like night, is inhospitable to vision; too many photons strike tree trunks before they reach a retina, too many animals have evolved coats that blend with the surroundings. I still only rarely see deer. Those animals continue to perk up their ears and twitch their noses to scan the air for information, for signals that move between and around stems and foliage; and still, they flee before I see them and before they see me.

I hear bird calls, but now wonder what mammalian conversations I might be missing. And I am struck by how odourless I find the forest. I watch dogs sniff about them as they follow the paths of previous visitors, and wonder what exactly those dogs – or deer, badgers, foxes or rabbits – know of who's passed by.

I shouldn't denigrate human hearing or olfaction too much. They are good, excellent even – a very solid mammalian inheritance. Also, I shouldn't celebrate our vision too much. Birds and other creatures see much better, although we put some serious brainpower to work on this sensory stream.* Most importantly, none of us should imagine that our senses present to us the world as it truly is. Our senses provide our nervous systems only with what is pertinent to our survival; to think otherwise would be as naive as a child believing the night belongs to only a few animals.

---

* Scientifically, it's very interesting to consider the importance of microscopes and telescopes, and how seeing the very small and the very big has been central to understanding the world better.

CHAPTER TWELVE
A Multilayered Brain Teaser

‘The brain of mammals, even the most stupid of them, has enlarged enormously,’ wrote Alfred Romer in 1933. ‘It is in the cerebral hemispheres, originally small structures dedicated to the sense of smell, that almost all the growth has taken place. Here there have arisen higher brain centres which have placed the mammals as a group far above any other vertebrate stock in their degree of mental development.’

Welcome to the first half of the twentieth century. Romer was a major authority on vertebrate biology, and he wrote this in his book *Man and the Vertebrates*. The idea that even the very dimmest mammal might be far brighter than any other vertebrate is quite something (and one assumes Romer didn’t think much of invertebrates either). This world view was typical of the period and asserted that the dawn of mammals had marked a quantum leap in cognitive ability. Mammals were making major strides on a path towards ever-greater intelligence that reached its zenith in *Homo sapiens*. It’s certainly a notion that would make a nice climax for a book written by a member of this most cerebral of species about what makes mammals so special – but is it true?

Unfortunately, no, it isn’t. It rings of outdated hubris. Things have become a lot more complicated since 1933. And brains are complicated anyway.

What is secure is that mammalian brains are enormously large. Compared to the brain of a reptile with a similarly sized body, a mammal has a brain six to ten times bigger. And, generally, larger brains make for smarter animals. But then, scaling brains to body size – as one must do to make fair comparisons between differently sized animals – one finds that bird brains are similarly inflated.

It’s also broadly true that most mammalian brains are large because of their bulky cerebral hemispheres – the folded grey matter that you see when beholding a human brain. However, this doesn't – as we'll see – tell the whole story of how mammals initially gained bigger brains. And the cerebral hemispheres are not simply overgrown structures originally concerned with smell.

However, the majority of the mammalian cerebral hemispheres consists of a neural tissue found nowhere else. And it's accounting for the origins of this tissue that lies at the heart of evolutionary neuroscience as it pertains to mammals. At stake is whether or not mammalian cerebrums truly contain a new type of 'higher brain centre'. And if they do, what the nature of this centre is.

As for mammals being 'far above any other vertebrate stock in their degree of mental development', this is what today sounds most snobbish. Mammals are clever. There's no doubt about this – it's a central aspect of what they are. But not all mammals are furry geniuses, and the more carefully people look at intelligence in non-mammalian animals, the more nuanced the situation is.

Darwin eventually discussed at great length how antecedents of human emotions and behaviours could be seen in other animals, but he mostly excluded brains from *On the Origin of Species*. It seems he wanted to build a case for his evolutionary theories without drawing attention to his belief that humans' music-making, poetry-writing, art-creating, science-doing, God-fearing minds were nothing but the products of souped-up monkey brains. After all, how could Victorian sophistication possibly have emerged from troupes of apes? And could the work of Darwin's contemporaries – from Dickens, Dostoyevsky and Eliot, through Browning and Whitman to Brahms and Liszt, Manet and Whistler – really be credited to a process that had chipped away at animal survival and reproduction?

Darwin also understood that biology was an aimless process. But as evolution became more broadly accepted, it was often viewed as a progressive – or even purposeful – process yielding ever better organisms, with humans as its masterpiece. Although this world view passed – the idea that human placentas were the peak of placental evolution was (fairly) quickly viewed as absurd – it cast its longest shadow over evolutionary neuroscience.

It wasn't just humans' intellectual and artistic achievements that skewed the discussion; by the mid-nineteenth century there were already over a billion people spread over this planet, and it was hard to pin this abundance on anything other than their brains. With technology and industry, agriculture and advanced everyday problem-solving skills, people had colonised a multitude of habitats and apparently become masters of their destinies. Humans were intelligent, and intelligence had served humans well. It was easy to view cleverness as a highly effective evolutionary strategy, and, vaingloriously, that higher cognitive ability was what evolution had always been building toward.

Neuroanatomy began to flourish only after Darwin's death in 1882. In its early days, researchers across Europe examined the cranial contents of a wide range of mammals and non-mammals. Some investigators attempted to uncover the microscopic arrangement of cells within these brains, while others examined the organs' gross anatomies. Soon, these studies indicated that the rear and middle parts of vertebrate brains were remarkably alike. Spinal cords, brainstems and hindbrains in fish, amphibians, reptiles, birds and mammals were distinguished by their similarities rather than by difference; in evolutionary terms, by conservation rather than divergence.

The rearmost areas of brains have more prosaic functions than the fronts. The brainstem, for example, regulates breathing and heartbeat, while other deep-lying centres execute ancient functions such as energy balance, or falling asleep and waking up. While an animal's different sensory streams feed into the middle and rear of its brain, the types of behaviours that these centres generate are rigid and instinctual.

Further forward, the midbrains and attached cerebellums of the assorted classes of vertebrate showed interesting variations, but it was at the front that the real evolutionary action had happened. What's more, one could line up vertebrate brains on an imagined scale from fish to amphibians to reptiles and on to mammals, and see an apparently progressive expansion of the forebrain.

In 1908, the German neuroanatomist Ludwig Edinger wrote, 'Finally, in mammals we meet a brain which has so large a [cerebral cortex] that we may well expect a subordination of reflexes and instincts to associative and intelligent actions.' The idea was that the higher – literally and figuratively – centres of the mammalian forebrain were capable of more advanced information processing and that they had taken over the functions of hindbrain centres in lower animals. What lower animals did instinctively, mammals did thoughtfully. Plus, if mammals were lined up just right, a progressive accumulation of cortex also accounted for the enlargement of brains as one moved upwards towards humans.

Then, in 1909, the Dutch neuroanatomist Ariëns Kappers introduced a whole raft of additions to the neuroanatomical lexicon. These terms sought to encapsulate the idea that brains had become increasingly complex along this linear scale by the back-to-front addition of new parts. Across different vertebrate classes, brain regions were given prefixes that indicated the apparent age of the structure in question. 'Archi-' and 'palaeo-' for old and oldest were attached to the front of the names of many fish, reptile and bird brain structures.[*] And the mammalian cerebral cortex was christened 'neocortex,' or 'new cortex'.

A cortex is any outer layer of tissue, the term being derived from the Latin word for bark. Neocortex is a thin sheet of neural tissue ranging from 0.5mm to 3mm (0.02–0.12in) in thickness across different mammals, which lies on the outside of the brain. Although the vast majority of this intricately folded sheet of grey matter coating a human brain – the very emblem of human intellect – is composed of neocortex, it grows out from the very front of the organ.

Human neocortex's characteristic crumples exist to enable more of it to be crammed into the skull – more than 75 per cent

---

[*] Archi- and palaeo- were, however, misused. Archi- should have meant oldest and palaeo-, old.

of a human brain consists of neocortex. However, cortex isn't massively enlarged in all mammals: hedgehogs, tenrecs and opossums have only small caps of it atop their brains. And while many mammalian cortices display the same space-saving contours as human ones, numerous species have entirely smooth ones. The cortices of monotremes are definitively mammalian, yet a platypus cortex is smooth, and an echidna's is all ridges and valleys.

Taking the right stain and a half-decent microscope to a sliver of neocortex – as, of course, you must, to begin to grasp its machinations – reveals that it is a structure made up of six layers running parallel to the brain surface, like sheets of paper stacked on top of one another. Each layer contains distinctive types of neurons, packed at characteristic densities, and wired to one another and to other brain regions in particular ways. And it is this six-layered structural arrangement that is seen nowhere beyond mammaldom, and that needs its origins explained.

We'll begin, though, with the slightly more tractable characteristic of mammal brains: their large size.

Jurassic spark

Few figures in neuroscience history are as maligned as Franz Joseph Gall. In the early nineteenth century, Gall concocted phrenology. He claimed that the bumps on a person's cranium could be inspected, and from them the size of the underlying brain parts inferred, and so, too, the person's mental attributes. This is plainly preposterous, but Gall should probably get more credit for helping to establish the centrality of the cerebral cortex to human thought, and for proposing that different parts of the cortex execute distinct cognitive functions. If you thought Darwin struggled to have his theories accepted, Gall's ideas were so materialistic and antireligious that they got him expelled from Austria in 1805.

While the idea that a brain's shape is apparent in the bumps on the outside of the skull may be silly, the underlying theory wasn't. Gall said that the larger a brain part was, the more highly developed its function was, and that is essentially true. And while phrenology is indefensible, reading the bumps on the *inside* of skulls has been pivotal to understanding the timing of mammalian brain evolution.

Plaster casts of the insides of fossilised craniums can be surprisingly rich sources of information about the organs those brain cases once contained. Most obviously, a cranium can reveal the overall size of its former tenant, but imprints of the brain's folds and fissures can also indicate the relative sizes of its various parts. At least, they can when the brain fills the entire cranium, as it does in mammals and birds. Unfortunately, the brains of most protomammals were held in cases within the skull, just as reptilian and amphibian brains are today.

To determine when exactly mammalian brains began their expansion required data from skulls of early mammals and their immediate ancestors. However, the further back in time you go, the rarer useful skulls become. And nobody is going to hand over their prize 220-million-year-old fossil head to let someone open it up and fill it with plaster. Luckily, you can now determine the internal morphology of a fossilised cranial cavity using X-rays, and in 2011 Tim Rowe, of the University of Texas, and colleagues did exactly this to chart the expansion of the mammalian brain.

A cast from a 260-million-year-old cynodont revealed that the animal's brain was small and tubular: the forebrain was 'narrow and featureless', and the midbrain not yet covered by cortex as it is in mammals. Rowe's opinion of the animal was withering: 'Compared with their living descendants, early cynodonts possessed low-resolution olfaction, poor vision, insensitive hearing, coarse tactile sensitivity, and unrefined motor coordination. Sensory-motor coordination commanded little cerebral territory.'

Next up was *Morganucodon*, the common early mammal known first from some Welsh teeth and bones deposited about 205 million years ago.* Rowe, however, had a

---

* I refer to *Morganucodon* and *Hadrocodium* as mammals – they each have the definitional jaw joint. Rowe – favouring mammals being

wonderfully complete skull found in China from which to assess its mental capacities. Relative to its 10cm (4in) body, the early mammal's brain was 50 per cent bigger than the cynodont's. It wasn't quite modern-mammal-sized, but it was getting there.

Most interestingly, the skull's shape said that the whole brain hadn't uniformly expanded. Instead, the most dramatic expansion had occurred in the smell-processing structures – both the olfactory bulbs (the first way station for olfactory signals) and the olfactory cortex (the second port of call) were much bigger. Additionally, the midbrain was now covered with cortex, which in this region is today concerned with processing touch information. And the cerebellum – a structure primarily concerned with coordinating movements – was enlarged too.

Hadrocodium lived about 10 million years after *Morganucodon* and was tiny. It was only about 3cm (1.5in) long and looked like a weasel that could sit comfortably on half your finger. *Hadrocodium*'s relative brain size indicated another pulse, coming in at the lower end of the range seen in living mammals. Again, the cerebellum was larger, and the olfactory regions expanded.

The functions to which an animal dedicates parts of its brain indicate what is important to that animal. That gaining better senses of smell and touch, plus a heightened degree of sensory-motor coordination, drove the initial expansion of mammalian brains fits with what other analyses have told us about early mammals. These nocturnal animals relied heavily on these two 'night-time senses' and they were gracile little creatures.

Brain regions concerned with hearing were not expanded in either *Morganucodon* or *Hadrocodium*. In *Morganucodon*, the mammalian middle-ear bones were still part of its jaw, whereas, interestingly, in *Hadrocodium*, the definitive mammalian middle ear existed, but the *inner* ear wasn't yet elaborated.

---

descendants of the last common ancestor of living mammals – calls them mammaliaforms.

These two animals were almost certainly hairy, and it might have been that sensory input from hairs drove – at least in part – the expansion of somatosensory cortex seen in *Morganucodon*. The presence of hair also says, of course, that these creatures were fully endothermic. Brains use a lot of energy. In terms of the organ's size, only warm-blooded birds match mammals for brainpower. The chronology provided by these skulls suggests that the initial phase of endothermy was not accompanied by a significant increase in brain size, whereas the second endothermy pulse, associated with insulation, was.

Knowing how diminutive early mammals were, and how endothermy today forces shrews – Mammalia's smallest living members – to endlessly scrabble for food, we can picture mammals 200 million years ago benefiting from sharper senses and more precise movement as they also continually sought sustenance.

In fact, people in the past have proposed that acquiring bigger brains was the primary driving force behind evolving endothermy. Today, it seems more likely that increasing brain size played into the feedback loop that sustained endothermy: greater energy acquisition permitted larger brains, while larger brains facilitated greater energy acquisition.

Earlier investigations of more recent fossilised mammal skulls had already shown that mammalian brains, on average, kept getting bigger, a trend that continued for tens of millions of years after the dinosaurs disappeared. This trend may sound like support for that ascending linear scale of intelligence that I was so quick to dismiss; however, the fossil record shows that this was no general increase. Across the various branches of the mammalian family tree, brains grew at different rates. The organ became largest in the primate radiation and the cetaceans. Scaled to body size, humans have the biggest brains, but many large mammals have more actual grey matter than we do. The sperm whale has the most of any animal, its nearly 8kg (17lb) brain weighing five times more than ours.

For many mammalian lineages, overall brain growth was due to the expansion of neocortex, but, as noted, numerous mammalian orders still contain members that have only small amounts of such cortex. A big brain or large cortex seemingly only evolves if it benefits the animal.

We'll return to cortical variability in living mammals, but now as we approach the problem of where neocortex came from, there's something I should mention. Olfactory cortex – whose expansion helped drive early brain enlargement – is not made up of six layers, it has only three; it is not a *neo*cortical structure. Given that the cerebellum isn't either, the initial brain expansion of mammals was not a neocortex-driven event.

This said, *Morganucodon*'s midbrain was covered with cortex that dealt with input from touch sensors, which probably did have six layers. For us to grasp the significance of having different layers of neurons stacked on top of one another, it's necessary to understand a little of how brains operate.

Circuit maps

There's an adage in detective work – at least in the TV shows I watch – that you've got to follow the money. In neuroscience, you have to follow the axons. Axons are the wiry projections that neurons send out to make contact with other neurons. It's along axons that action potentials – brief spikes of electricity – travel, and it's in the pattern of these spikes that information is carried from A to B.

At B, the spike causes neurotransmitter to be released from the axon's terminals, and this transmitter either increases or decreases the chances of the next neuron in the chain firing more spikes. Understanding which neurons are connected to which other neurons provides a circuit diagram of a nervous system from which to infer how the system processes information. That's why vast amounts of time, effort and money are currently being spent drawing ever finer-scale maps of brains.

The simplest type of circuit is a reflex arc. Say you reach out to touch something you don't know is burning hot. You touch it. And before you know it your arm has been flung back, your elbow is folded tight, and your hand is on your chest. I say 'before you know it' not just to indicate the speed of this response, but because it's true. This reflexive action happens independently of your brain – signals only arrive up there to make you exclaim 'Ouch!' after your hand's been removed.

Your hand withdraws thanks to a unidirectional three-stop circuit consisting of sensory neurons → interneurons → motor neurons (see Figure 12.1). The sensory neurons have axons that run from your fingertips to your spinal cord, where they meet the interneurons. If a shock of heat sends action potentials racing up these sensory axons, then a spritz of neurotransmitter excites the interneurons. Then the interneurons likewise spike and chemically stimulate the motor neurons. Motor neuron axons exit the spinal cord and run to muscles in your arm and hand. Spikes travelling along these axons instigate the release of neurotransmitter onto the muscles, causing them to contract, and your hand to dart back.

Many animals operate solely, or primarily, according to hard-wired reflexive responses to stimuli they commonly encounter. But the neocortex is, of course, considerably more complex than this circuit. Nevertheless, let me make a few points. First, neuroscientists use the pattern of spiking in neurons to examine how they encode information – for the aforementioned sensory axons, the hotter the touched object is, the more they fire. Hence, temperature is apparent in the

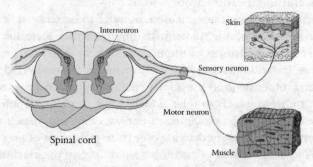

Figure 12.1: A reflex arc – one of nature's simplest neural circuits.

firing rate, while functionally, more closely spaced action potentials more quickly excite the interneurons.

Second, this simple circuit illustrates one of the core functions of a nervous system. Without a body-wide network of neurons, a finger with a heat sensor could mount only a local response to being hot – perhaps the finger muscles themselves could contract. But by the sensory neurons passing the 'FINGER HOT!' signal to the central nervous system, a far more efficient and widespread response is generated. A coordinated contraction of various muscles means the entire arm mediates a more complex withdrawal motion. The nervous system has taken a very specific input and commanded an appropriate response at the level of the whole organism. Similarly, a wildebeest throws its whole body back from the riverbank when it sees a crocodile; it doesn't just blink.

Third, you can picture how adding extra elements to the circuit can alter its function. Say, for example, a thread of axons ran down from the brain to terminate on the spinal interneurons, and that they could *inhibit* the interneurons. An addition like this would allow the reflex to be turned off if necessary; when, perhaps, the brain deemed it beneficial to briefly grab something very hot and move it. So, extra circuit elements increase the circuit's flexibility, and consequently enable the animal to generate different behaviours according to circumstance.

And fourth, remember earlier in Chapter Eight talking about the transformative power of parenthood and how hormones can change behaviour? Well, hormone receptors – and those for other neuroactive messengers – more subtly alter the way neurons act. If an animal senses it's in danger, it typically releases a surge of adrenaline. Adrenaline doesn't activate reflexes directly, but if it acts on receptors on certain neurons, it can make those neurons more or less *prone* to fire should another neuron send messages to say they should fire. Hypothetically, an adrenaline-primed interneuron could fire in response to three sensory-neuron action potentials rather than the usual, say, five. Hence, if you're expecting to be ambushed, you might jump at a slight touch rather than wait for a fuller bite-force. Or a wary wildebeest might startle at a floating log, rather than waiting to fully compute if it's a log or a crocodilian predator.

In this reflex arc circuit of just three types of neuron, you can see: (1) information coding; (2) the animal mounting a complex and beneficial response to a simple stimulus; (3) a basis (with a fourth node added to the circuit) for alternative behavioural responses to the same stimulus; and (4) how the hormonal state of an animal can change its behaviour. These are some of the reasons nervous systems are useful. And they illustrate how brains can soon get very complicated.

To approach neocortex, sensory axons delivering a 'FINGER HOT!' signal to the spinal cord don't just kick-start a reflex. They also activate neurons that relay the information up the vertebral column to the brain. The spikes that travel there make you *know* you just hurt yourself.*

And with the signal in the brain, interesting things can happen. The 'hot finger' message will arrive at the same time as streams of information from other sensory systems, and timing will bind these inputs together. From the eyes, there will be a visual description of *what* was hot, and a report that it was sitting above some blue flame-like things

---

* I have no idea why you *feel* hurt. I purposefully haven't mentioned 'consciousness', which, from both a neurobiological and an evolutionary perspective, is undoubtedly fascinating and in need of explanation. The phenomenon hovers over the whole of neuroscience. But you can study all the stuff I'm talking about – neurons, action potentials, synapses, information coding – and discuss it in objective terms that make perfect sense in terms of biology and never need to discuss that, in parallel, nervous systems create an enigmatic mental phenomenon by which we *experience* these events. So, yes, when we burn ourselves, we *feel* hurt. Yet the physical analysis of neuroscience is entirely compatible with consciousness not existing. How do most practising neuroscientists deal with this problem? Typically – professionally, at least – they ignore consciousness. And that, I'm afraid, is exactly what I'll be doing here.

on a big metal box-like structure. The reflex arc responded only to hot, but in the circuits of the brain, information from multiple sensory streams will interact. Following the axons, and the spikes they carry, one sees how visual and bodily information first travels to areas of neocortex dedicated to a single sensory system, then later collide in different areas where the higher-level thought 'saucepan was hot' can be realised. The areas in which this type of amalgamation occurs are called 'association' areas, and they too reside in the neocortex.

Neocortex also detects patterns in sensory information. Take vision and the richness of detail we perceive with that sense. It's amazing that to open your eyes is to have a visual scene appear instantaneously, with no vestige of its creation in multiple brain regions. But, for sight as good as ours, a great deal of neural territory, executing a great many computations, is involved: numerous brain circuits extract different features from the retinal activity that a scattering of photons generates, and stitch them together. Human brains have separate cortical areas and neurons for processing colour and movement, assessing depth, detecting edges. And the brain imposes constancy on the inputs it receives; it has the wherewithal to keep track of a single crow flying across our view when in reality all that exists is a series of retinal images with different black shapes on them. Adding brain areas not only allows more behavioural flexibility, it also allows more sophisticated interpretation of the flow of sensory information.

Finally, the brain stores memories of its experiences, so that the animal might learn from its past. In mammalian brains, a structure called the hippocampus – which has its obvious homologues in reptilian brains – is essential for creating and storing memories of events. With any luck, you will recall picking up a hot saucepan from a stove and not repeat the mistake.

The social learning we discussed in Chapter Eight, which reaches its zenith in humans, means that animals can avoid grabbing dangerously hot saucepans if someone tells them not to, and forgo flirting with predators if they've seen their cousin mauled by one. These are complex phenomena, requiring that the implications of another's behaviour can shape an individual's future behaviour. So, more circuits, more computations. We humans can hear the words 'Don't touch things on lit stoves', and process the abstract concepts conveyed by words, store that advice and relate it to real-world objects to direct future actions. Sure, you might, if you're curious, carefully extend a hand to sense the warmth two inches from the pan of boiling water, but you won't just grab it.

A nervous system that can generate multiple behaviours, and that can pick the most appropriate one at a given instance, makes an animal a responsive and flexible organism. A nervous system that perceives the external world in great detail and that can make more sense of complex fluxes of photons or sound waves lets an animal respond to a wider range of signals. And a nervous system that learns makes an organism that might be – should be – better adapted to its world the longer it lives. While natural selection sifts the products of random genetic change to preserve organisms best equipped to function in their surroundings, the evolution of advanced nervous systems has yielded organisms that can themselves adjust to their environments; systems that can adapt in seconds, minutes or hours, rather than over millennia.

Neocortical circuits

Just as the late nineteenth century saw certain dyes transform our understanding of chromosomes, at the same time new staining techniques also allowed pioneering neuroscientists, such as Santiago Ramón Y Cajal in Barcelona, to decipher the microscopic organisation of the brain. One revolutionary stain labelled at random just a handful of the thousands of neurons in a thin slice of brain but picked out those few cells completely. The stain revealed the neurons' entire glorious morphologies: axons could be seen shooting off one way and dendrites – the spindly branches that extend from a neuron's cell body to receive axonal inputs – were observed branching into their individual trees. And, when Cajal looked very closely, he saw that there were

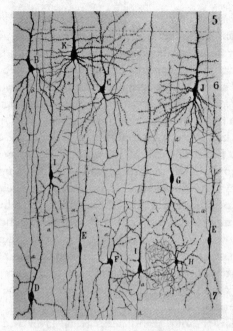

Figure 12.2: From 1887, Santiago Ramón y Cajal exquisitely drew the brain's microscopic structure; here, an array of cortical neurons from 1904. Source: Science History Images/Alamy Stock Photo.

tiny gaps between the axons and dendrites: synapses, the minute gaps across which, we now know, neurotransmitters diffuse.*

Another dye found in late Victorian times coloured neuronal cell bodies purple and, in staining all neurons, it was critical in revealing the layers of the neocortex. It took some time for researchers to agree that there were six, in part because different areas of cortex vary in their microscopic organisation. This variability was confirmed and brought to fruition in 1909 by the German anatomist Korbinian Brodmann. Brodmann looked across the entire human neocortex and described the exact nature of the layers in different places. In distinct regions, the layers fluctuated in terms of their relative thicknesses,

---

*Very much my *second* favourite experience of standing gobsmacked before an open notebook in a museum happened when I encountered a page of Cajal's incredible handmade drawings of neurons.

whether or not they were divided into sublayers, and how exactly their constituent neurons were arranged. Altogether Brodmann counted about 50 areas in humans, producing a 'cortical map'. If you imagine lifting the sheet of neocortex off the brain's surface and laying it out flat, such a map resembles a country divided into counties or states (see Figure 12.3). Brodmann called the areas the 'organs of the brain', proposing that each had its own discrete function. Working a century after Gall, Brodmann could point to work by neurologists showing how injuries to particular areas of the cortex affected specific cognitive functions. He suggested that the variations in cellular organisation acted as the substrates for achieving the unique computations that mediated each area's function. The paper is now a classic. Nevertheless, across the different areas Brodmann described, the cortex seemed to be coming up with variations on a theme, as if nature were subtly jamming on a basic – or canonical – circuit constructed across the six layers that characterise all neocortex.

Layer 1 is the outermost neocortical layer and contains very few neuronal cell bodies; instead, axons run through

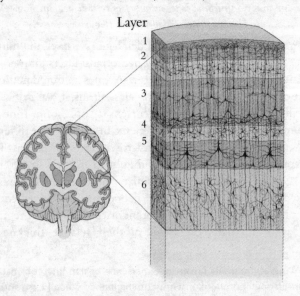

Figure 12.3: Mammals' cerebral cortices contain six distinctive layers.

this layer, contacting the tops of the dendrites of other layers' neurons. Below this is Layer 2/3 (although it's always called six-layered cortex, splitting Layers 2 and 3 appears to be for aficionados and they're typically lumped together). Layer 2/3 contains neurons that lie at the heart of the neocortical circuit. Then, in Layer 4 there are smaller neurons, which receive inputs from the axons of a brain region called the thalamus, the thalamus being a structure that conveys most sensory information to the cortex. Layers 5 and 6 each contain fewer but larger neurons, whose axons exit the cortex.

The stereotyped connections between these layers and the way specific layers connect with the brain beyond each cortical region mean that there's an essential route by which information flows through neocortex. The realities of the cortex's many connections are complex, but the take-home message is that the layers of neocortex constitute a variable but elemental circuit whereby input from thalamus → Layer 4 → Layer 2/3 → Layers 5 and 6 → output target, another brain region (the nature of that target being dependent on the region under consideration).[*]

Once you understand this circuit, disparities in the ways different cortical areas are constructed make more sense. For instance, primary visual cortex has a much-expanded Layer 4 to receive information from the eyes, whereas primary motor cortex – whose axons head off to instigate the movements of muscles – barely has a Layer 4 but has a huge Layer 5.

If this is the basic circuit, we can finally ask, where did neocortex come from? While the short answer is no one's really sure, we'll explore the two main theories.

---

[*] All the neurons I've spoken about are excitatory neurons, whose spikes and neurotransmitters press the next neuron into action. However, interspersed between the excitatory neurons are many different inhibitory neurons, whose transmitters act to dampen activity. These cells are critical for filtering information, for curtailing runaway activity – such as that seen in epilepsy – and for controlling the timing of activity.

Neocortex as new

With brains, we're back to issues arising from mammals being
the only surviving members on our fork of the amniote family
tree, that is from there being nothing in between a reptile and
a mammal. A problem that's especially challenging when a
mammalian trait is fully formed in platypuses and echidnas, as
neocortex is. Strictly, the best we can say about neocortex is
that it evolved at some point between mammals' and reptiles'
last common ancestor and the split between monotremes and
therians. If it was indeed present over the midbrain of
Morganucodon, we could narrow its range of birthdates to
sometime between 310 and 205 million years ago.

As for what neocortex evolved from – the equivalent,
say, of exactly which sweat glands turned into mammary
glands – the best we can do is look at the brains of other
living amniotes and try to infer a possible ancestral state.
Then, from there, we must hypothesise *how* neocortex
evolved. The large brains of birds have clearly expanded
and changed in their own particular ways, but perhaps
other reptile brains resemble older ones. A turtle, for
instance, has three types of cortex, each one relatively
simple. There is an olfactory cortex on the outer side of its
brain, which is very much like mammalian olfactory
cortex – it too has three layers, with only the middle one
containing an abundance of excitatory neurons. On the
medial side of the turtle forebrain is another simple cortex
considered homologous to the mammalian hippocampus.
Again, this has a monolayer of neurons – these functioning
in memory formation – flanked above and below by two
other layers. Then, on top of the turtle brain, there is a
small cap of dorsal cortex, likewise made up of a single
layer of excitatory neurons with a few inhibitory neurons
scattered about. It processes visual and somatosensory
information.

Now, if you look at a hedgehog or an opossum – one of
Mammalia's less cortically endowed constituents – the
arrangement of olfactory cortex, neocortex and hippocampus

isn't too dissimilar to this basic set-up. Theory One of the genesis of neocortex has it that reptilian dorsal cortex transformed into neocortex, with the single layer of neurons gradually multiplying.

In the reptilian dorsal cortex, the single population of excitatory neurons acts as both input and output layer; that is, the same neurons receive information from the thalamus, then send their spikes down axons back out of the cortex – there's no real intrinsic circuitry for information to whizz around. The evolution of neocortex from this structure is, therefore, a process of replacing a single, multi-tasking neuron with a circuit of four or five specialists. It's a bit like replacing a general housekeeper with a dedicated cleaner, butler, cook and maid (if you like your analogies a little *Downton Abbey*). As progressively more elements were added, the functions of cortex were divided between the additional relays in a circuit. And as previously discussed, adding extra cells and synapses would have increased the possibility of being able to execute further computations.

Because no one has ever found a cortex intermediate between reptilian dorsal cortex and neocortex, the order in which layers were successively added, and when, are matters of speculation. Would sail-backed *Dimetrodon* have had two layers of neurons, say – one receiving inputs, the other sending outputs? Did therapsids have three-neuron cortical circuits? We don't know.

Comparisons of three- and six-layered cortices have shown that the interactions between excitatory neurons and neighbouring inhibitory neurons are similar in the two types, providing a basic framework for incorporating new neurons. It's thought that in mammals the developmental process of generating new neurons has been extended to produce more neurons, and that successive neuronal generations pop off the production line in slightly different forms, so as to produce specialist Layer 4 neurons, say, or those of Layer 2/3.

An additional pertinent difference between reptilian cortex and neocortex is the orientation of axons from thalamus; in reptiles, they run parallel to the brain's surface and contact many cortical neurons, whereas in mammals they shoot up from the base of the cortex contacting fewer neurons more powerfully, thus increasing sensory resolution.

As dorsal cortex only processes information from reptilian eyes and touch receptors – hearing remains a subcortical sensation – the expansion of neocortex, once it got a foothold, involved the development of new areas, as Brodmann described, each assuming a new function.

In this schema, mammalian ancestors were like electrical engineers or computer scientists, developing a new type of brain circuit: the neocortical microcircuit. In the second model, they were no such thing. They were more like rather creative architects.

Neocortex as not so new

Ana Calabrese did her neuroscience PhD in a statistician's lab; like what happened to evolutionary biology in the middle of the 20th century, brain functions these days are increasingly being expressed in mathematical equations. Calabrese was interested in generating computer models of how neurons responded to sensory input. Such models tend to start with a known set of inputs and a known set of neuronal responses, and then try to figure out what neural circuits do – in computational terms – so that different neurons fire in their characteristic manners.

Generating such a model is usually enough to earn a PhD, but Calabrese decided she'd like to work on data she'd gathered herself from real-life neurons. To do so, she crossed Columbia University's Morningside Heights campus to work in a lab that studied birdsong in zebra finches. Together with the lab's professor, Sarah Woolley, Calabrese inserted arrays of tiny electrodes into three different regions of the finches' brains and recorded the responses of 823 individual neurons to sounds akin to those that constitute the birds' songs. Calabrese and Woolley were interested in how bird brains worked, not in how they'd evolved.

After Calabrese had examined the resultant mass of data, which showed that neurons in various parts of the birds' brains fired at distinct times and in different ways, she presented her analysis at a conference. There, the person most taken with her observations was an expert not in birds but in mammalian neocortex. Ken Harris, of University College London, saw that for nearly every type of avian neuronal firing that Calabrese had classified, there was an equivalent neuronal subtype that spiked the same way in the cortex of mammals.

Harris suggested that when Woolley and Calabrese published their study, he would write an accompanying commentary. When the paper came out in 2015, Harris explicitly compared three bird–brain regions to separate layers of mammalian cortex. Bird brains were not supposed to act like mammalian brains. Unlike the layered cortices of mammals, the avian forebrain is a collection of blobby nuclei, entirely different in overt anatomy. But, Harris emphasised, avian neurons in 'field L2' acted like mammalian Layer 4 neurons, firing first and coding information in a similar manner. In both mammals and birds, these two regions are the respective first targets of axons from the thalamus. Then, neurons in another bird nucleus acted like Layer 2/3 neurons. Neurons in these two structures fired later and in a more complex manner, but again the parallels were striking. And finally, mammalian Layer 5 neurons had their equivalents in the bird brain's 'field L3', albeit in the four criteria that Harris discussed these neurons corresponded on only three, rather than all four as in the previous pairs. The conclusion was stark: operationally, the circuits in these two distantly related types of animal, with their outwardly dissimilar brains, were strikingly alike. Harris wrote, 'If there is a canonical cortical microcircuit, and if this circuit is indeed homologous between birds and mammals, it means that this circuit was operating in the last common ancestor of mammals and birds over 300 million [years] ago.'

This radical idea wasn't new, however. Harris, Calabrese and Woolley saw that these new observations supported a theory that had been around for nearly half a century. In the

1960s, American neurobiologist Harvey Karten proposed a theory positing that everyone had been misled into thinking that the nucleated brains of birds and the laminated brains of mammals were more different than they really are.

When Karten decided to investigate the machinations of avian brains, birds still had the reputation of being dimwits. The surveys of the various brains of vertebrates at the opening of the twentieth century had been unsympathetic to birds. The fact that their forebrains contained nuclei rather than layers had been the main problem.

The nuclei were deemed to mark the brains as primitive. They looked like what lies *beneath* the mammalian neocortex: the subcortical mass of nuclei with stripes of nerve fibres running through them that we call the striatum. The striatum has always been associated with simpler, more fixed patterns of behaviour (in contrast to the deliberative and intelligent cortex). Based on their superficial similarities, birds' forebrains were judged to be nothing but hypertrophied striatums, capable only of producing behaviour much less flexible than mammals' forebrains, even though, gram-for-warm-blooded-gram, birds had brains that had expanded to the same degree as mammals'. As late as the 1970s, Alfred Romer kept dismissing birds as feathered automata with limited mental capacities.

Karten joined the avian field, however, just as this central idea underlying birds' lowly cognitive status was being challenged. He would contribute both to demolishing it, and then to building a new conception of the avian brain.

Mammalian striatum isn't only nucleated and striated, it also contains characteristic neurotransmitters and enzymes, and has very well-defined connections with other brain regions – all of which identify it far more concretely than its outward appearance. And when researchers began to look for these striatal signatures in birds, they didn't see the whole forebrain light up. They found, instead, that just one small region of it resembled the mammalian striatum. In fact, the ratio of striatum-like to non-striatum-like forebrain was strikingly mammal-like. This left the obvious question: what did the rest of the avian forebrain do?

Karten's approach was to map the brain's connections and draw its circuits. Starting with the sensory systems, he carefully followed the routes that auditory, visual and somatosensory information took. Systematically, he traced the pathways taken by information from the ears, eyes and spinal cord, as it passed from one relay station to the next through the avian brain. Periodically, he says, he got very excited about a new pathway he'd found, only to discover that when he presented it to his colleagues at the Walter Reed Army Institute of Research in Maryland, they'd say, 'That's exactly the same as it is in mammals!'

Eventually, Karten saw so many parallels between the circuits of avian and mammalian brains that he proposed a theory. In 1969, he wrote that the circuits weren't just similar, they were the same – bird and mammal cognition operated on a common basic circuit that had been conserved for over 310 million years.

Karten says it's like looking at a Californian Modern house versus a tall, thin New York townhouse – they look radically different, but they're still built to provide their occupants with a kitchen, bedroom, bathroom and living space. The basic elements – the neurons that speak to each other – are the same, they're just arranged very differently. In this model, the long independent histories of the mammalian and avian lineages have built very different brains, but their core circuits have been preserved. As Harris echoed, the core circuits on which nucleated bird and laminated mammal brains worked had been present in their last shared ancestor. In 1969, the theory was left field and iconoclastic. And, Karten says, 'it went over like a fart in church.'

You can see why. The general perception of birds remained one of feathered dullards, people were only just getting to grips with their forebrains not being giant striatums, and here's a guy saying bird brains work just as the splendiferous mammalian neocortex does. Frequently, Karten was dismissed as trying to cram avian brains into a mammal–shaped mould. But, he says, he was just reporting the data.

Karten subsequently went off to work on retinas, cataloguing the variety of cells they contain. His focus became understanding visual pathways. But he occasionally updated this evolutionary work. Generally, though, the theory sat on the fringes of evolutionary neurobiology in a box marked 'Interesting if True (Bit Far-fetched)'. Most people merely accepted that neocortex had arisen from reptilian dorsal cortex.

Calabrese's work, however, is one of a clutch of recent studies that have given Karten's hypothesis renewed vigour. In 2010, Karten himself, now in San Diego, published observations from chicken brains indicating that they contained circuits more like the precise organisation of mammalian cortex than ever previously seen. Karten's key message was that to look for homology between nervous systems you have to look at the level of neurons and circuits. Their arrangement in layers or blobs – or any other configuration – is secondary.

The alternative to similarities between avian and mammalian brains indicating that their core circuits are a shared inheritance from a common ancestor is that birds and mammals convergently evolved overlapping ways of performing neuronal computations. Perhaps the strongest argument against this possibility comes from recent research that has shown how genetic markers that identify specific mammalian cortical neurons during their development are also expressed by immature bird neurons that will grow to have comparable functions in the adult brain. This is proposed to be a molecular remnant of their shared origins. Studies published in 2018 extended the known genetic similarities between mammals' and birds' neurons, as well as showing these likenesses extended to lizard, turtle and alligator neurons too. This expanding dataset is now allowing researchers to estimate which genes drove neural development in the last common ancestor of amniotes.

Why mammalian ancestry ended up organising the elements of this circuit in a layered sheet of cortex, while in birds they formed a nucleated brain, remains an open question. And *how* exactly mammalian forebears reconfigured the architecture of the ancestral circuit into layers is one of the largest challenges to the hypothesis. For Karten to be right, the dominant theory of how neocortex develops would have to be wrong, although Karten believes some studies have actually observed an alternative form of neuronal migration that means his model is plausible and that models of cortical generation *should* be modified.

Because of this issue and some others, numerous people think Karten is simply wrong. For them, the theory still carries a bad smell. The other chief objection is the claim that the embryonic tissue that develops into the nuclei containing these circuits in birds can be identified in mammals, and that there it develops into a smaller brain structure that is very much not neocortex. Asked how to reconcile these viewpoints, Karten, a staunch experimentalist, speaks plainly: 'Don't try and reconcile, collect more data.' The molecular studies published in 2018 should please him.

The most arresting element of the Karten hypothesis is the idea that birds and mammals use core operating systems that were invented 310+ million years ago. To understand when evolution might have concocted the circuitry seen across amniotes, a greater characterisation of the circuitry of reptilian, amphibian and fish brains is required. Karten says there are provocative findings accruing in these organisms; now, these nervous systems should no longer be dismissed as simple, and entirely different from mammals' and birds'. For example, recent studies of lampreys – representatives of the earliest, 'simplest' vertebrates whose lineage branched off from jawed fish 450 million years ago – show that their striatums, at least, are startlingly like mammals'. And investigations of fish brains are rewarding neuroanatomists with exciting new insights. If Karten is right, nervous systems – the very neural networks that evolved to allow animals to be flexible – would be, in their core features, remarkably conservative.

Looking back in evolutionary time, Karten acknowledges the enormity of the questions being posed. 'What is the origin of this whole story? When did it really begin? That's the exciting challenge, that's such an exciting question to ponder. This is not a challenge that has an answer, it's a challenge that has more and more questions with every answer you get.'

Brain evolution was driven, like all evolution, by changes that allowed animals to survive better in an uncertain world.

What a profoundly human privilege it is to exercise our neocortices on trying to understand their origins.

Big brains redux

At the end of his consideration of Ana Calabrese's study, Ken Harris wrote, 'Perhaps intelligence isn't such a hard trick after all: a basic circuit capable in principle of supporting advanced cognition might have evolved hundreds of millions of years ago, but only adapted to this purpose when the benefits actually outweighed the costs of increased head size, development time, and energy use.' That's quite a departure from the triumphant 1933 sentiments of Alfred Romer.

In requiring large heads and extended periods of development, and in consuming copious amounts of calories and oxygen, large brains capable of higher intelligence come with significant costs. Whether an animal evolves towards greater intelligence is therefore a question of costs and benefits. A favourite anecdote illustrating such trade-offs charts the life history of the sea squirt. As larvae, these animals are like tadpoles with tails, eyes and nervous systems that propel them across the ocean. But as adults, they sit rooted to the seabed and feed on passing plankton – a mode of life that requires little intelligence. After the sea squirt has chosen where to spend its adult life, it undergoes a metamorphosis that involves digesting its own eyes and nervous system!

Brains evolve in specific biological contexts. Perhaps no bird has written a symphony because needing to fly negates the growth of a truly large brain. Or maybe animals are constrained by their inability to have opposable thumbs. The dextrous hands (and feet) of primates allow the world to be finely manipulated so that thoughts can be turned into action. Whereas the large brains of dolphins and whales, and their unquestionable intelligence, evolved in bodies with much more limited abilities to handle and change the environment.

What contemporary behavioural neurobiology – the science by which we infer animals' cognitive abilities – is finally getting wise to is the need to more sympathetically and carefully observe and test various animals' behaviour if we are to understand their intelligence.

Transforming our conceptions of bird brains has run in parallel to a greater appreciation of their intelligence. For years, Irene Pepperberg was considered something of a crank for spending her days talking to a parrot called Alex. Sure, parrots could talk, but these animals were thought to just copy – *i.e.* parrot – human behaviour. Under intense scrutiny, however, Pepperberg showed that Alex had developed a vocabulary he understood, that he comprehended the concept of colour and could do simple arithmetic. Crows, too, can use simple tools, and in Japan, they even use cars to crack open nuts. These birds drop nuts onto pedestrian crossings and let cars drive over them. Then, when the lights turn red, they swoop down and feed.

Most studies of rodent behaviour employ rats and mice who've been food-deprived, so that they'll run mazes and perform mini intelligence tests for food rewards. When someone tried this with alligators in the 1960s, they concluded that alligators were stupid. But then we've already discussed how infrequently crocodiles – alligators' closest cousins – eat. When the alligators were offered the opportunity to choose different temperatures at which to rest as a reward for performing tricks, instead of food, they showed they were quite capable of the tasks asked of them.

In recent decades, neuroscience has retreated from examining a wide array of organisms. Much research now focuses on rodents and primates with an eye on understanding the human brain. The old assumption of a linear scale running from rat to monkey to human can feel lazily implicit. In fact, common experimental species are called 'model organisms', rather than being viewed as animals in their own right, as products of their own unique evolutionary course. Recently, however, some researchers (often those who work on species not typically found in labs) have become more vociferous in calling for a stronger evolutionary perspective. Their argument seems sound; to look across many organisms is to see how brains vary, to see what is essential and which elements evolution has modified to tweak cognitive function and behaviour to different species' advantages.

Finally, if the rewards of intelligence appear to be writ large in the current existence of seven billion human beings, know that our species has teetered on the brink of extinction more than once. Only 70,000 years ago there may have been fewer than 10,000 people alive. Only with the advent of farming, and advances in basic technologies – inventions a long time coming, but that could then be socially transmitted – did *Homo sapiens* span the globe. And only later still did they find themselves comfortable enough, and cumulatively smart enough, to build neuroscience institutes.

Neocortex redux

Brodmann's 1909 identification of multiple different areas nested within the vast expanse of human cortex is one of neuroscience's seminal insights. Regardless of the exact origins of neocortex, Brodmann's work likely holds the secret of how the neocortex has underpinned the many and varied cognitive adventures different mammal lineages have embarked on.

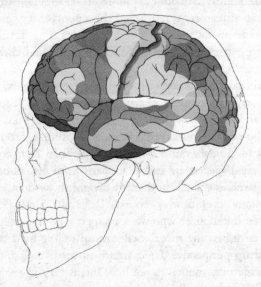

Figure 12.4: The mammalian cortex is divided into anatomically distinct areas that carry out different functions.

The variations in the cortical layers that Brodmann saw are now interpreted in terms of how a canonical neocortical circuit is specialised in each area. Always, the circuit's conserved building blocks are recognisable, but their exact arrangements are distinct enough to indicate that brain evolution updates this circuit independently in each region. Harris – with Gordon Shepherd, of Chicago's Northwestern University – has called the cortical areas 'serial homologs'. We might then liken them to fingers: each of those five structures is clearly a variation of the same structure, yet each is unique, and their complementarity is what makes a hand a success.

Additionally, with each area having a degree of autonomy, it can form its own set of external connections, whether these are with other cortical regions or with targets below the cortex. Together, these two facets – internal structure and external connectivity – make each cortical area its own functional unit carrying out a particular task. And fascinatingly, that job can vary from being the deciphering of sensory information (of any modality) to controlling the body's muscles or mediating more abstract cognition, what we might call *thinking*.

'Cortical areas are a fantastic way of increasing what you can do with cortex,' says Jon Kaas, of Vanderbilt University, Tennessee, 'they make for a wonderfully flexible brain.'

Kaas has studied extensively how different species of mammals have different collections of cortical areas, reflective of their lifestyles. Such surveys compound the notion of a conserved structure that can profitably diversify. In the small cortices of bats, for example, there are prominent auditory areas attuned to the sound frequencies these animals use to echolocate. Rodents' auditory cortices are expanded in the regions that deal with the high frequencies of their calls, and the platypus bill feeds cortical areas for electrical sensation. In nearly blind subterranean mammals, the visual cortices have regressed to leave only the smallest remnants. This adaptability evokes other mammalian innovations. Like the mammary gland, which has evolved to deliver an enormous diversity of lactational regimes, and the development of an endothermic physiology that sustains camels and arctic hares, pygmy shrews and blue whales, in neocortex, mammals again invented something exceptionally malleable.

Kaas has also asked how many distinct areas the cortices of the first mammals probably had. By comparing the brains of monotremes, marsupials and representatives of the four major branches of placental mammals, and focusing on which areas are common to all or most mammals, Kaas estimates that early mammals had 20 areas or fewer. He thinks somatosensory areas would have been large components of early cortex, which is consistent with Rowe's fossils. And all living mammals have some visual and auditory cortex, plus areas that process information arriving from muscles that help establish where in space the limbs are. The first mammals also probably had some association areas, which combined information from the sensory areas.

Interestingly, Kaas proposes that early mammals did *not* have dedicated motor cortices. Instead, the outputs that ran to voluntary muscles arose directly from sensory areas. Although not everyone agrees, Kaas says most marsupials still lack specific motor cortices, and that such structures are an innovation associated with placental mammals. How finely, one wonders, must you examine the movements of an opossum versus a rat to see the functional consequences of this area's emergence?

Successfully coordinating two motor cortices on either side of the brain – each controlling the muscles of the opposite side of the body – has been mooted as the driving force behind the evolution of another uniquely placental mammal innovation: the corpus callosum, a highway of axons that connects the left and right cortices of these animals. Many other distinctive features of mammalian brains are the results of subcortical regions becoming more extensively wired to neocortical regions.

A small – perhaps a square centimetre or so – 20-area cortex was, therefore, the starting point for mammalian brain evolution. In most lineages the brain has expanded, although in the tiny brains of shrews there doesn't appear to be enough room for all the usual areas. A cortical area below a certain size wouldn't have the neural firepower to operate, and so shrews have jettisoned some dispensable areas to keep the essential ones in working order.

Conversely, when neocortex expands, two things can happen: either each area expands proportionately, or more areas appear. Although new areas can pop up to serve niche needs – such as echolocation or electrosensing – they also permit more sophisticated computation. 'The principle,' says Kaas, 'is that a larger number of areas can serially process. Like a computer, you can come out with remarkable things if you do things over a number of steps. You do a simple computation and repeat it and repeat it, and come up with amazing results. That's why the human brain is so amazing.'

We now know that Brodmann rather underestimated the human cortex. We don't have around 50 cortical areas, but about 200. Many believe this is where the key to human intelligence lies. To take vision as an example, sensory input can arrive in one area, be processed there, and the end result passed on to another area for further consideration. A vast expanse of humans' cortex is dedicated to visual perception, and it is a potpourri of distinct processors extracting different features from what the retina delivers to the initial relays.

In addition, the association areas – where information from multiple cortical regions mixes – are enlarged in the human brain, especially at the brain's very front. The prefrontal cortex integrates information received from all over the cortex. Remarkably, this region doesn't mature fully until a person is in their twenties. Now in my sixth year of parenthood – watching my older daughter grapple with reading and writing, and the social circus of the school playground, and listening astounded to the exponential growth of my younger daughter's vocabulary – I see we are only a fraction of the way to the maturation of this organ.

One recent winter afternoon, I sat with my lunch in a London park under the city's grey sky and had my attention stolen by a squirrel in the naked canopy above. It traversed the tree tops – branch to branchlet to twig, one tree to the next – as if it were skipping across solid ground. When one of its mates shot down a tree trunk, I watched the creature gather a nut of some sort in its forepaws and nibble for a while. Briefly, it paused. Stood upright. Scanned for danger, then bolted straight back up the trunk's vertical face.

What a torrent of information the squirrel brain must process as it flits around its arboreal home. How graceful and fully controlled its movements are. How unlike me it is.

In watching the squirrel – climbing, eating, scanning, skipping, and above me nearly flying – it seemed obvious that its brain would have been shaped by the demands of its life. Its brain served its body. How funny that we humans have flipped this viewpoint on its head, that we see our bodies as our brains' servants.

CHAPTER THIRTEEN

This Mammalian Life

Cristina and I made a photo album of Isabella's first year. It's the most joyous book you'll ever see, a testament to vitality, growth and love. But its pages are laden with subtext. Someone looking through it might notice that the man who beams as he holds Isabella alongside Cristina's mum near the beginning never reappears. And that person might question too why only my mum and brother appear on my family's first visit to New York. If they do, they might appreciate more my dad's elation when he later holds his granddaughter aloft.

That year, even after we'd left behind the violent emotional swings of the NICU, illness and wellness never ceased to press on us from either side. While my brother is being an uncle for the first time, and my mum a grandmother, my dad is convalescing from chemotherapy. When we are in England that summer, Cristina's stepfather lies in a coma. When we visit in November, it is for his funeral.

Throughout the year, we all kept on asking, hadn't we been through enough in the NICU? We each wanted to plead that we'd learnt enough there about chance, fragility and the spectre of mortality. Apparently, we hadn't.

Dad, today, is in remission. He's happiest playing with his grandchildren. Isabella and Mariana have been elemental in sustaining both him and Cristina's mother in their respective recoveries from illness and bereavement.

If a misdirected football hadn't struck me in a uniquely mammalian attribute, this book wouldn't exist. It wouldn't exist either if a meteor hadn't struck the earth in the Gulf of Mexico 66 million years ago. Life hangs on unlikely events. But only in fatherhood did the project take shape. My being a mammal felt initially like a somewhat arbitrary biological identity to consider – the result of yet another man absorbed by the iconography of a baby at its mother's breast, perhaps. Then, I was unaware of what a pivotal position parenting would occupy in this story, of how quintessential maternal care is to mammalian life. Did I allow myself to get swept up in that aspect? Maybe a little – another person's take on being a mammal might have spent much longer on the sophistication of the mammalian kidney; people have written entire tomes on the marvel that is mammalian dentition ... Those accounts exist, should your appetite be whetted.

Looking back now, I see a project entirely infused with the idea of birth. Each chapter questioned how new traits were born and what new possibilities arrived with them. I was also drawn to the birth of new knowledge and new ideas, to the varying routes by which biology comes to eventually better grasp its subject matter. And, of course, a central concern became the birth of mammals as a group, as a unique type of animal.

But in any account of evolution, mortality lurks on every page. Flora and fauna change by the differential progression of germlines; perishable bodies are always cast aside, and the ratio of death to reproduction is what's elemental. The succession of pelycosaurs, therapsids, cynodonts and mammaliaforms is as much about extinction as creation. What consistently struck me, surveying that sequence, was how tight the margins are: when a palaeontologist labours over incremental changes in shoulder joints, tooth shapes or jaw-bone morphologies, one sees how it all matters, how natural selection – over vast spans of time – is sensitive to the tiniest of differences. How delicate the balance between survival and extinction must be if teeth can be shaped by the speed with which they extract insect innards; if a slightly more excessive secretion of milk's forerunner pushed one lineage of animals forward; and if how an ear that heard fractionally quieter or higher-pitched sounds made all the difference.

Also, at this project's very beginning, I felt I was likely embarking on a straightforward celebration of a superior type of animal. In these accounts of the traits that define mammals,

I believed I'd see what made mammals better. This notion seemed well supported by people calling the post-dinosaurian epoch the *Age of Mammals*.

I haven't at all ceased to marvel at the ingenuity of mammalian biology – quite the contrary – but peering at its component parts hasn't always yielded the anticipated uniqueness or superiority. The prevalence of live birth beyond mammalia is an obvious example, while the parallel evolution of so many traits in mammals and birds was a constant theme. The three-boned middle ear of mammals is beautiful, and unique, but other animals have independently developed excellent hearing. Lactation is definitional, but some birds have their crop milk, and there are fish whose skin secretes mucus that their young consume. The idea that mammals as a group are all the smartest critters out there, we now see was not a smart one.

Even the idea that we live in the *Age of Mammals* can feel like a throwback to older times when evolution was viewed as inherently pushing forward to higher ideals, and to the making of *Homo sapiens*. Since the dinosaurs departed, their avian descendants have proliferated widely, and their reptilian cousins have persisted too, with twice as many species as there are of mammals. In the seas, more than 26,000 species of ray-finned fish swim. Plus, a million species of insect exist and hundreds of thousands of other invertebrates. The world is split into many niches accommodating microorganisms, plants and animals and their vastly different lifestyles. Mammalian-ness is only one answer to the question of how to live. What makes it interesting is not that it is better, but that it is ours.[*]

Humans alone disregard these bounded niches – our brains have allowed us spill out all over this planet. We farm, we hunt, we consume, and we now do so wastefully. I write these words within weeks of reports that cheetah populations

---

[*] It's telling that therapsids displaced pelycosaurs, and were replaced by cynodonts, who, in turn, were victims of mammals' success, indicating that they competed with one another to occupy the same ecological niches, while other vertebrates lived alongside them.

are rapidly shrinking, that giraffes must now be considered vulnerable, and that 60 per cent of primates are threatened with extinction. The grim report on the precariousness of so many primates also described how feel-good reports about newly discovered species of primates should be read with caution – new species are often encountered as humans expand their activities deeper and more destructively into these animals' homelands. Cheetahs, giraffes and most of our closest relatives join an already alarming number of species teetering on the brink. The idea that a future student of mammals might pick up the tenth edition, say, of *Mammal Species of the World: A Taxonomic and Geographic Reference* and place on the table only a lightweight book is heart-breaking.

The pages of *MSW* will not shrink for some time, for the book includes any mammal that's lived in the last 500 years. The rationale behind that is that half a millennium is no time at all on a geological scale and that maybe, just maybe, species feared extinct live on unseen. In 1962, a Russian whaling ship may have spotted some Steller's sea cows, animals it seemed European hunters had wiped out in 1768, just 27 years after first encountering them. Nobody will declare extinct a species of bat found on a New Zealand island, but not seen since the 1960s after rats arrived via a shipwreck; a microphone may have detected their calls in 2009. But hope is not enough to counteract the grim effects we – the brightest but most selfish of mammals – are having on this planet.

I, primate

It is the ancientness of every species that makes the loss of any species tragic. An organism is a cumulative thing.
I, tetrapod.
  I, amniote.
    Then *I, pelycosaur.*
      I, therapsid.
        I, cynodont.
          I, mammal and finally, *I, placental mammal.*

That has been the story of this book: these Russian dolls of biological identity.

To step closer to our brilliant, if careless, human selves, molecular genetics' overhaul of the mammalian phylogeny said that next comes *I, Euarchontoglire*. That's where we left things in Chapter Nine. (I'm confident I picked the catchiest book title out of this lot.)

The next historical split was the division of Euarchontoglires into ancestors of rabbits and rodents – collectively Glires – and those of Euarchonta. Euarchonta consists of primates – of which *MSW* lists 376 species of lemurs, tarsiers, monkeys and apes – plus two species of colugos and 20 or so species of Scandentia. Scandentia are known as tree shrews, but they're not true shrews, and some live on the ground. Colugos are called flying lemurs, but they're not lemurs and they can't fly; they do, though, look fairly lemur-ish, and using a sheet of skin linking their front and rear legs they glide long distances between trees.

Trees are the glue that links Euarchonta together. The founders of this group were small, nocturnal tree-dwellers that ate insects, and most closely resembled the living tree shrews. Only a few of these animals' descendants have ceased to live above ground.

The first primates are known from 55-million-year-old rocks,[*] and these creatures had evolved two essential traits – eyes in the fronts of their heads, and grasping hands and feet. Having forward-facing eyes where the two visual fields substantially overlap allowed for the development of 3-D vision – the brain comparing parallel inputs to create a richer perception of depth. A useful trait for a life spent moving through tree-tops, although you see it too in predators who must judge depth to kill. Unlike many other animals found in trees, primates don't remain around the trunks and lower limbs, they live high in the canopy. There, they have hit upon a rich diet of fruit, buds and insects. A 2017 study suggests that the nutritious value of fruit especially may have helped

---

[*] Naturally, their origin is estimated to be much older by molecular dates …

spur the evolution of larger brains in primates. And as one might expect of highly visual animals, primates have come to be mainly active in the day. All monkeys, apes and humans – except the imaginatively named owl monkey – are diurnal.*

Primates' grasping hands and feet have on their fingers and toes specialised pads that increase their grip and heighten their sense of touch. And they have nails instead of claws – although many lemurs, tarsiers and a few monkeys retain a singled clawed finger on each hand to aid in grooming. Their centres of gravity have shifted towards their hindlimbs, which dominate locomotion, giving them a unique way of walking.

Overall, the primate radiation is an exploration of arboreal movement and of vision: a lifestyle that has led to profound neural transformations. Fossilised craniums show that even the earliest primates had brains large for their body size, the expansion driven principally by enlarged brain regions that processed visual input. The fine-scale inspection of living brains required to truly make sense of this organ reveals several additional specialisations. Whereas in most large-brained mammals the neurons themselves are bigger too and more spread out, in primates the neurons remain the same size and are packed at the same high density, meaning larger primate brains have many more neurons. Additionally, the primate thalamus – the brain's first major relay for most sensory input – harbours unique processing centres for both vision and the bodily senses.

However, it's again the cerebral cortex that is most interesting. The visual cortex is not only large in primates, it is also divided into more and more definable sub-regions. Certain areas mix sharp eyesight and finessed muscle control to function in visually guided movement. Primates evolved expanded regions of cortex for receiving detailed sensory information from the hands, and regions – apparently unique – dedicated to movement *planning*. Finally, they possess enlarged prefrontal cortices – regions at the front of

* Most mammals that are active at night have small eyes, indicative of their greater reliance on other senses, whereas owl monkeys and nocturnal lemurs have enlarged eyes.

the brain where multiple strands of information are combined and analysed. While detailed cortical maps are available for only a limited number of mammals, it nevertheless appears that primates have many more functionally specialised regions than other mammals.

I don't wish to present a linear ascent through ever-smarter primates: bright mammals emerged on various prongs of the mammalian family tree, and not every primate is an arboreal Einstein. But undoubtedly many of these cerebral changes were essential for the very particular intelligence with which we humans are familiar.

Another general feature of primate life is that things slowed down. Primates have relatively long lives, they are late to reach sexual maturity and, typically, they reproduce only one baby at a time. That baby has a long gestation and normally emerges as a well-developed young mammal before continuing on its rather slow course of growth.

Finally, primates tend to be very sociable. Most of them live in groups with complex interactions. It all sounds rather familiar, doesn't it?

About 30 to 25 million years ago, apes emerged from the primate clan. Then one day, maybe six million years ago, some of these apes started to move on only their hindlimbs, limbs that would eventually carry a particular lineage – and everything they'd learnt in the trees – out onto the savannah. There, full coats of hair would become redundant, childhood would extend even further, and their brains would grow to unprecedented dimensions ... No, stop, that's a different book. But those singular animals moving to a human future surrendered none of the characteristics that made them mammals; rather, the upright and cerebral lives of *Homo sapiens* depend on the traits that define mammals.

I, mammal

I thought at one point my big take-home on mammals was going to be about how the traits that define them are so wondrously adaptable. Lactation is capable of quadrupling the

size of an already hefty newborn hooded seal in just four days;
it can sustain a litter of ten piglets and form the basis of an
eight-year bond between an orangutan and her single
offspring. Endothermy permits bears to live in the Arctic and
camels to traverse vast deserts. Mammalian teeth can grind
goodness from grass, seeds and insects or rip gazelle meat to
pieces. Mammalian brains pilot flying insectivores that
hunt their prey with sonar, lead humpback whales on
16,000-kilometre (10,000-mile) migrations, guide the relatively
straightforward life of a hedgehog, or, yes, conjure Bach's
Cello Suites and devise theories of evolution. All of this is true
– the malleability of these traits is profound, the variety and
wide-ranging distribution of mammals unquestionable – but
I'm not sure it's specific enough to mammals. After all, if
there are more than a million species of insects out there,
insect parts must be dazzlingly adaptable. Instead, what's
important is the *combination* of individual traits.

It was the most natural thing in the world to map out this
book in chapters that each dealt with an individual aspect of the
mammalian body. This Linnaean-type approach not only
chimed with what you usually see when you look up what a
mammal is – 'a warm-blooded vertebrate with fur and mammary
glands' – it echoed the division of universities into departments
and the partitioning of hospitals into wards that specialise in
healing particular body parts. Biologists have long studied
organisms by breaking them into their constituent parts.

But I knew before I even commenced the book that the
chapters could not all stand alone. The origins of the scrotum
didn't lie in the autonomous biology of sperm production. To
explain why most male mammals produce their gametes
beyond the sanctuary of their abdomens, one had to consider
the way mammals had either become endothermic or had
benefited from a new mode of locomotion. Lactation, too,
was a trait bound to the changing biology of mammalian
ancestors: when the eggs of our forebears became smaller and
warmer, they ran the risk of either drying out or being
overrun by microorganisms parasitically fuelling themselves
on exuded thermal energy.

Mammals were clearly becoming increasingly warm-blooded, yet here were two examples of roadblocks that endothermy encountered along the way. Picture a lineage of animals moving towards progressively higher body temperatures. And say they have all the physiology to achieve this, all the usual things you associate with a high-energy lifestyle – strong heart, lungs and jaws; gracile limbs; intricate teeth – then, bang, their eggs can't cope. They're drying out, they're rife with microorganisms. Well, *that* didn't seem like an endothermy issue. Natural selection acts to address this problem, and a vaguely milk-like secretion evolves. Then 100 million years later, bodies get warmer still, and they start to gallop around as ever more agile creatures, then, bang, sperm production becomes an issue. Who'd have seen *that* as a barrier? The linkages between body parts or physiological processes are not always straightforwardly predictable.

So, early on, I took a sheet of A3 paper and crudely sketched these connections. And the further I progressed, the more lines needed to be added. If middle-ear bones were ancestrally jaw bones, a line between eating and hearing was required. The new ears fed information to enlarged brains, which consumed enormous energy budgets – more lines. Elaborately shaped teeth were useless without jaws and musculature that moved them appropriately – a little triangle of intersections. It was impossible to ring-fence any trait. No chapter – as I trust has become apparent – could be an island.

It wasn't that I expected this matrix to have escaped people's attention, but still, there was a strange shock of recognition when I first encountered Tom Kemp's diagrammatic take on mammalian biology. In his rendition, 30 different nodes were linked by an intricate web of lines. By comparison, my drawing looked rather childish.

For Kemp, there's a grave problem with dividing a mammal – or any organism – into constituent parts. Yes, individual traits can be meaningfully defined and studied in isolation, but such an approach should never blind us to the fact that every trait is a part of a greater whole. And it is the whole organism – the mammal, fish, tree or bacterium – that must survive and reproduce.

Functional interactions between different body parts are pivotal to understanding the evolution of *any* type of organism. The interactions create *interdependences*, and therefore no single attribute – no 'key innovation' – can be credited with defining, or making possible, a group of organisms.

Which brings me back to another expectation I had at this project's inception: I thought one trait would emerge as the defining point of mammalian life. The three that have historically had the strongest cases made for them are endothermy, intelligence and lactation. I confess to having been most enthralled by the transformative impact of lactation. But while there is no dispute that radical innovations open new ways of life, Kemp's point is that lactation, for example, did not evolve in isolation.

St George Mivart was concerned with how something as sophisticated as a mammary gland could evolve, and very detailed analyses of genetics and development were required to answer his question. But, in addition to the mechanics of transforming a sweat gland into a milk dispenser, the animal hosting this reconfiguration had to be one capable of amassing great amounts of food, and of storing surplus energy in a form that could later be converted into dairy products. She had to be caring for her eggs in the first place. Without the right starting condition and parallel adaptations, lactation was impossible.

And just as the emergence of lactation depended on these multiple factors, as soon as mothers fed their offspring milk, this mode of feeding entered, and reshaped, a new web of interactions – many of them reciprocal. The energy bill for milk production is high, but by allowing the evolution of fully occluding teeth that grew in an animal's adult-sized jaw, lactation helped make energy-gathering more efficient. Extra energy could feed bigger brains, which made for better hunters or quicker-witted herbivores. These brains could more sophisticatedly process information arriving from ears that owed their acuity to the expansion of the jaw in the first place. On and on it went, in loops and complex interdependences. You simply can't pick a trait and say, 'There! That was the key one!'

The idea that key innovations carry decisive leaps forward is akin to me giving Mariana a canvas, paints and brush and expecting her to be able to reproduce van Gogh's *Sunflowers*. It isn't going to happen – she and those things are mismatched. Instead, we provide her with materials appropriate for her current abilities, but materials that stretch those abilities as far as they might extend at this point in her development. Her artistry will develop as her mind, muscle control and the materials she's given inch forward together, integrated with each other.

As Kemp puts it, 'There is no identifiable single key adaptation or innovation of mammals because each and every one of the processes and structures is an essential part of the organism's organisation.'

How then to explain the evolution of complex organisms? Kemp talks about 'correlated progression', the essence of which we touched upon in his views on the evolution of endothermy: any trait in a complex organism is only free to change a small amount before further change is required elsewhere.

Kemp's favoured metaphor is to liken an organism to a line of people who walk forward hand in hand. Each person is an individual trait, and for the line to remain a viable entity – for the organism to remain alive and functional – no pair of hands can come unclasped. The woman standing for teeth can take a stride forward at a certain point, but for the whole line to move, the man representing the jaw needs to take a step forward before Ms Dentition can make further progress.

The other key point Kemp makes is that functional interactions are not fixed. My examples of egg health and spermatogenesis hampering endothermy were transient interactions – impermanent clasping of hands. Once the scrotum existed, the challenges of endothermy moved away from the testicles, and once viviparity evolved, endothermy was likely advantageous for embryo development rather than being the problem it had become to egg-laying mammals.

One sees that the current nature of any organism defines its possibilities, defines how and where it might evolve.

Kemp is now officially retired from research but continues to write. He is charming and erudite, and I regret immediately confessing my inability to remember species' Latin names when I meet him at St John's College, Oxford. Over lunch, he explains that he's able to dismiss the notion of 'key innovations' to his undergraduates by making the case that all mammalian biology evolved for the benefit of the big toe.

After we've eaten, I ask him if he can offer a definition of a mammal. He recalls the first of his webbed diagrams, published in 1982. At the centre of that diagram, he placed the word 'HOMEOSTASIS' and around it, three main nodes: 'Temperature Regulation', 'Spatial Control' and 'Chemical Regulation'.

Homeostasis − meaning the maintenance of constant conditions − is a term coined in 1926 by Walter Cannon, but a concept expounded as early as 1865 by the French biologist Claude Bernard, who said that the 'fixity of the internal milieu is the condition of the free life'. As such it applies to all life: inside any organism local order is upheld against the disarray of the outside world. But Kemp sees the ability to maintain homeostasis reaching its zenith in two types of animals: birds and mammals.

When tetrapods emerged from their ancestral aquatic homes, the challenge of sustaining an internal milieu founded on water-based biochemistry was immense. At the animal's borders, water-filled cells met air; gone was the medium of water with its inherent buoyancy, and ambient temperatures could fluctuate wildly across days or across seasons. The multitude of adaptive changes that saw the first land animals transform over 100 million years into the first mammals meant each of those challenges was met. Mammals operate at constant body temperatures, have bodies that efficiently traverse the unevenness of terrestrial landscapes, and they maintain their internal chemistry accordingly. The homeostatic strategies mammals developed are beyond the capabilities of a newborn, and so each generation of mammals ushers in the next. In fact, a womb might be seen as a means by which a mother extends the constancy of her physiology to her developing young, while her fat storage and provision of milk shield the young mammal from potentially erratic food availability.

It's perhaps ironic that whales, manatees and seals have returned to aquatic ways of life, but testament, too, to the adaptability of this core plan, its ability to sustain life in such a variety of habitats.

This focus on homeostasis and Bernard's remark on freedom make me think of a comment by J. Z. Young in his famous 1950 book *The Life of Vertebrates*. 'The camel and the man he is carrying through the desert may perhaps contain more water than is to be found in the air and sandy wastes for miles around,' wrote Young. 'This is only an extreme example of the "improbability" of mammalian life, which is one of its most characteristic features.'

Throughout our conversation, Kemp is convivial but serious. Only towards the end, sitting outside beside two long-empty coffee cups, do I see his posture fully unfurl as he leans back, smiles and says, 'I think mammals are wonderful things.'

Endnote

I arrive at Down House – Charles Darwin's bucolic Kentish home from 1842 until his death in 1882 – in search of a scientist; a one-man scientific institution. Darwin wrote his books here, and did his experiments in laboratories housed in these grounds. It is a crisp blue autumn day.

The ground floor of Down House uses a mixture of original and substitute possessions to recreate the house as it was when the Darwins lived there. The upstairs rooms – bar the restored marital bedroom where you learn that Emma Darwin read novels to her husband at the end of each day – are more typical of a conventional museum, and it is here the visitor is encouraged to begin. In the first bedroom, a short video plays on a loop, with some talking heads saying that Charles was a family man. One even states that his work came second to his children.

After rooms dealing with the Galapagos, we are back in England again. Among masses of scientific artefacts are dotted quotes from family members. Darwin's son Francis reports that his father's desk evoked 'an air of simpleness, makeshift and oddness'. This is what I want, insights into the genius. Then, soon after, there's a curious 'Darwin Family Poem':

Write a letter, write a letter
Good advice will make us better
Father, mother, sister, brother
Let us all advise each other.

Next, in a room in which photos of the children are displayed, words from a granddaughter recall how shadows of the leaves of the old mulberry tree outside the window – still there, but now propped up – once danced about the white floor. Emma and Charles had ten children, seven of whom reached adulthood; two died as babies, Annie aged ten. Around the corner stands an old wooden slide that, hooked onto the stairs here, amused two generations of children.

In the master bedroom, I stand and behold Darwin's morning view of the English countryside. How strange to think that for all that he witnessed while circumnavigating the globe, this was the gentle landscape he saw most often. Descending the stairs, I ponder whether to go straight to his study, or whether to visit the rooms as they present themselves and save the study until last.

I begin in the drawing room, and listen as the audio guide describes how Emma Darwin played the piano for her family every evening. The room is large but intimate; the mismatched furniture lived-in; board games are scattered about the place. Darwin used to put jars containing earthworms near and on the piano to watch if these animals could hear or sense vibrations. He was, it's often remarked, not a distant or disciplinarian Victorian father, but only here in his home are you gently, progressively enwrapped in the routines of his family life. The air seems to hold traces of the flux of children, the bustle of seven or eight vivacious youngsters, and the fondness that passed between the members of this family.

Darwin's study – where *On the Origin of Species* and the tomes that followed it were written – is square in the middle of this home. To stand by the window that lit Darwin's personal microscope and to peer at his bookshelves, desk and piled-up miscellanea is exhilarating. But now, the imagined scene is reconfigured; there would have been little silence here, Darwin's intense cogitation would have been punctuated by his children's attentions, his work proceeding against a backdrop of familial thrum.

For lunch, I sit in a corner of the café and take out the book I brought with me: a collection of Darwin's letters and memoirs compiled with Francis Darwin's *Reminiscences*. I dip immediately into the son's recollections. Randomly beginning with holidays, where Darwin would seek out interesting plants and determine how local flowers were fertilised, Francis writes, 'My father had the power of giving to these summer holidays a charm which was strongly felt by all his family.'

But then he is quickly on to his father's persistent ill health, and from there to the loss of Annie. Ten-year-old Annie's death is well known to have devastated Darwin and is often credited with having destroyed the final vestiges of his Christian faith. As such, it is frequently discussed in terms of its alleged effects: the disregard with which a godless universe treated a pious man is viewed as having finally freed Darwin to publish the revolutionary theories he'd kept quiet for years.

Francis quotes at length a passage Darwin wrote within days of Annie's death as a safeguard against dimming memories:

> Her joyousness and animal spirits radiated from her whole countenance, and rendered every movement elastic and full of life and vigour. It was delightful and cheerful to behold her … When going round the Sand-Walk with me, although I walked fast, yet she often used to go before, pirouetting in the most elegant way, her dear face bright all the time with the sweetest smiles … In the last short illness, her conduct in simple truth was angelic … When I gave her some water,

she said 'I quite thank you;' and these I believe, were the last precious words ever addressed by her dear lips to me. We have lost the joy of the household, and the solace of our old age. She must have known how we loved her. Oh, that she could now know how deeply, how tenderly, we do still and shall ever love her dear joyous face!

Certain expressions evoke Mariana, in others I see Isabella, then ultimately there is only Annie. In New York, in the Museum of Natural History, having looked at Darwin's epochal notebooks, I was moved by Annie's death but swept along by the greater scientific narrative. I sit, now, plunged into 160-year-old torment, and think, I'd rather you'd had your daughter, we could have waited for everything else.[*]

The Sand Walk that Annie pirouetted around was Darwin's 'thinking path', a pebble and sand track around a pocket of woodland at the back of Down House's gardens. For much of his life, Darwin walked it three times a day, on each visit doing five laps to complete a one-mile walk. To count, he kept a pile of flint stones at the Sand Walk's first corner and kicked one aside each time round. To walk this path was always my priority in coming here.

As I approach, passing the greenhouse where Darwin examined a multitude of different plants and the kitchen garden where different varieties of vegetable were compared, the audio guide explains that the Sand Walk too was no place

---

[*] This is how I experience this passage. However, science historians John van Wyhe and Mark Pallen have recently cast serious doubt on the widely propagated idea that Annie's death dealt the final blow to Darwin's piety. As far as they can determine, this theory was only formulated a century after Darwin's death and is based on conjecture and very indirect evidence. Somehow, it has caught the public imagination, but van Wyhe and Pallen present convincing evidence that Darwin had reasoned himself out of Christian faith much earlier – probably before Annie's *birth*. They do not question the anguish Annie's death caused Darwin.

of solitary contemplation. While Darwin walked, the children played in the middle; occasionally, they'd replace a kicked flint to make their father encircle them an extra time.

Today, I am alone in the woods. The ground is laden with fallen leaves, and the path crunches with scattered beechnuts. Decaying wood lies propped against the living. The steely hard greens of holly shine against neighbouring autumnal hues. Trees, shrubs, ferns and groundcover nudge at one another. Each plant is so obviously seeking advantage, looking to survive, looking to produce another generation. Darwin too must have trodden on this seed fall – these trees' thousandfold attempts to sire new offspring – knowing how infrequently a young tree broke the old canopy. Everything he sought to explain was here.

Two squirrels join me. Grey ones, of course. I'll later try to figure out whether Darwin ever watched squirrels here. Grey squirrels were only introduced to England in the 1870s, but were there red ones at Down House? Eventually, I revisit Francis's reminiscences and learn how his father was forever stealthily creeping about or standing dead still in these grounds to look for 'birds and beasts'. And that once 'some young squirrels ran up his back and legs, while their mother barked at them in an agony from in the tree.'

I stop to watch this grey pair. They care very little for me. But when one retreats, the other pauses and fleetingly gives me his attention. His tail looks like a question mark.

We are no camel with a human rider in the midst of a bone-dry desert, but we feel an unlikely pair. Two furnaces of relentless metabolism free of our surroundings. Two random endpoints of unknowable spans of history. And I think what a very tiny fraction of the life in this forest we constitute, how very many more plants there are than mammals; what luxury items we two are.

The squirrel and I stand in a space where one of the brightest, most endlessly curious intellects nature ever forged turned its attention to Nature herself to yield a theory of why things are as they are. The squirrel skips on. His tail echoes the arcing movement of his body – a feathery *m* hops towards the tree. Without pause or apparent thought, he transitions from flat ground to the sheer face of the trunk.

He ascends and disappears, and I feel suddenly flushed with gratitude – that it's my very good fortune to be here. I decide to walk another lap.

Afterword: Mammals 2018

On a bright spring Sunday, six months after this book was first published, I took my morning coffee and sat on a bench peering into old English woodland. Before me the sub-seasons were shifting, early spring yielding to late. The bluebells – no longer magic – covered the forest floor still, but now only as yellow-green clumps of leaves apologetically propping up their seed-heads. To watch spring unfurl, that day, one had to peer upward, upward into a canopy shimmering in virginal greens, greens too delicate, too pure to last.

Most of these leaves sprang from bustling hornbeams and birches – these sprightly youngsters hustling for light around and beneath two old oaks only just starting to reawaken, seemingly content to let everything around them get a head start. Elsewhere shrubs, grasses and scrub clamoured into leaf. Everything wanted to grow now – where the sun shone, green raced to be.

Across the gulley where a winter stream occasionally flows, a sagging wire was pinned to flagging wooden posts. There to mark a border, a demarcation of what was ours (my mother-in-law's) and theirs – an offstage Lord to whom great swathes of this forest belong – it seemed that day absurd. I wondered what the wood's inhabitants made of it. It was too high to trouble a badger and too low to present a deer with a serious hurdle. For birds, it was just another perch. Maybe, they barely minded it at all. But, still, there it was, a metallic reminder that humans owned this place.

Not that *Homo sapiens* conjured the concept of ownership out of nothing – many species mark and maintain territories and ranges. And wire is undoubtedly preferable to Lord What's-his-name arriving here weekly to cock his leg against a tree. What felt preposterous was how minimally attached to this land its owner is – does he come here at all? What is this place to him?

Then, on cue, a movement. To my left, just beyond the wire, a ruffle of bluebell remains. I expected to see a squirrel but instead saw a fox. *No, a fox cub; no, two fox cubs. Hang on… three! What? No. Let me count these properly – one, two, three, four – no that's the mother – four, five… Is that? Yes, it is – six! That can't be right.* But it was. I counted again. Six cubs and their mother. *Oh my!*

By a clump of hornbeam trunks, the six youngsters played in an arc around their mother, who lay on a mound looking out. They pawed at objects, did little jumps and turned brief rolls as if they were enacting an advert broadcasting the importance of playful activities to growing mammals. And as a reminder, too, that play can be a social or individual behaviour, four cubs played solo, while two entertained each other.

Then, watching how the mother's presence allowed these new animals to develop whatever skills they were developing, I recalled how an essential union of successive generations of mammals had emerged as a central theme of this book.

Indeed, I thought constantly about how much my daughters would enjoy seeing this, but was certain that if I got up to leave, the foxes would flee, and that if they didn't, they most assuredly would when the girls approached. No, though – after I'd tiptoed away, I managed to marshal Isabella and Mariana to the forest's edge, and with Cristina and her mum behind, a family of five watched a family of seven, elevated by spring and life and all its vitality.

* * *

In 2018, scientists published more than a quarter of a million studies tagged with the keyword 'mammal'. The interactions between generations of mammals were nicely highlighted in a study that showed how North American moose and bighorn sheep learn their migration routes from their elders. In Germany, researchers showed how, in an apparent bid to reduce running costs, shrews' brains shrink for the winter and regrow in the spring. Researchers at the University of Cambridge grew something like a placenta in a dish.

Meanwhile, sequencing the entire genomes of Himalayan marmots and koalas, respectively, gave insights into how a mammal goes about living at very high altitudes or subsisting entirely on a diet of eucalyptus leaves. Koalas apparently have more taste receptors for detecting plant toxins than other mammals, presumably to allow them to distinguish good eucalyptus from bad. Also down under, scientists fathomed how wombats defecate cuboid poos – these animals use piles of their turds to signal to each other, and apparently this shape grants the piles more stability while the wombats are stacking. It turns out the last 8 per cent of a wombat's intestines have regions of varying elasticity that cleverly sculpt this shape. And, finally, on the subject of poo, in Alaska, it was found that fruit-eating bears are a major dispersal mechanism for seeds and that, consequently, many small mammals scramble to gain access to the rich delicacy that is bear scat.

Then, in the category of papers that undermined mammals' uniqueness, researchers in Germany and Italy found that the Somalian blind cavefish – an animal that swims in complete, subterranean darkness – has lost a molecular safety system that protects DNA from UV damage, and it's the same one that placental mammals lost during their long nocturnal stint. In China, scientists found that certain ant-mimicking jumping spiders – and, seriously, these creatures are dead-ringers for ants – do something remarkably akin to lactating, feeding their hatchlings a nutritious secreted solution, albeit from their egg-laying orifices rather than from nipples. *Ah, life.*

But oh, death. The foxes I watched in May were seven of an estimated 357,000 currently residing on the British Isles according to the UK's Mammal Society. This summer, the society published a survey that assessed the 'population size, geographical range, temporal trends and future prospects' of all 58 species of terrestrial mammals that live wild in the UK. Temporal trends were determined by comparing this year's population sizes to those that had been estimated at the last such census in 1995.

Water voles, wild cats, black rats and certain bats are all critically endangered on these shores and many other species, including the beloved hedgehog, are in sharp decline.

At the launch of the report on the survey's findings, a journalist asked whether mammals might have an image problem – contrasting the popularity of birdwatching to an apparent paucity of mammal-watchers, the term itself alien. Surely this was untrue, everyone said, but no one could really give a concrete rebuttal. I realised later that the UK's three most common mammals – field voles, moles and wood mice, all of which are roughly as numerous as people here – are animals I've never seen in the wild.

The way that many mammals live their lives out of view – of humans and of as many other animals as possible – contributed to the survey's other defining feature: its high degree of uncertainty. The Mammal Society reckoned that there are 6.3 million pygmy shrews in the UK but, it said, there could be as few as 999,000 or as many as 38.9 million. The population of Daubenton's bats was estimated at a healthy 1.03 million, but more realistically it is somewhere between a glowing 4.44 million and a distressing 27,000. (The problem with bats is that many people are able to report that a mammal just flew by, but few can say to which species it belonged.) Part of me liked this mystery and the way it implied that even today, on these most domesticated of islands, mammals can be genuinely wild. But if we're to truly know the fates of the mammals that live among us, better methods and greater precision are needed.

There was uncertainty, too, in a study out of Israel and California that sought to calculate the mass of all living things on this planet, an endeavour that confirmed my suspicions in Darwin's garden that there really is much more plant than mammal in this world. Globally, for every one gram of wild mammal, there are 65kg of plant, which is to say that for every ton of plant there is, there's just half an ounce of mammal.

Plants account for 82 per cent of earthly life's total weight. Next comes bacteria (about 13 per cent), then fungi, followed by the other single-celled groups, archaea and protists, before animals come in second from last, ahead of viruses only. Although, with mammals constituting only 10 per cent of animal life and 0.03 per cent of all life, there's actually less mammal than there is virus on planet Earth.

If these numbers are surprising, it's because the reality of ecological balance is surprising. But the way the weight of mammals is distributed these days is startling because humans are quite something. *Homo sapiens* – that tiny twig on the mammalian branch of life – now constitute about a third of all mammalian biomass. And, if you throw in our mammalian livestock – mainly cattle and pigs – together we represent 96 per cent of all mammalian biomass. All of this world's five-and-a-half-thousand species of mammal living untamed are outweighed 24-to-one by humans and their domesticates.

Drawing these conclusions is technically challenging, the authors said, and deriving historical trends is harder still, but the researchers cite a study that estimates that since modern humans have been around, the total mass of wild mammals has fallen by about 83 per cent.

According to another 2018 study, when humans are around it is the largest mammals that are the most vulnerable. Here, authors from a number of US universities used the mammalian fossil record since *Homo sapiens* and our immediate ancestors have existed to relate a mammal's body size to its chances of going extinct. The researchers observed that over the last 125,000 years, the bigger a mammal was, the more likely it was to have bitten the dust. Notably, this wasn't the case for the preceding 65 million years, and so this change is almost certainly the result of hominin activities.

All those giant sloths, giant deer, cow-sized armadillos and mammoths whose spectacular skeletons populate natural history museums – and which make visitors strain to imagine how, when and where such behemoths roamed – were victims of the way feelings of awe and a rallying call to slaughter are conjoined in us. (Another report this year described the discovery of ancient human footprints nested inside those of a giant sloth, a fossilised record of our species stalking its prey.) Projecting just 200 years into the future, the authors saw nothing in current population trends for elephants and rhinos, hippos, giraffes and other large terrestrial mammals to suggest they'd still be around. By then, they suggested, cows may well be the largest land mammals left alive.

Not only are humans making mammals, on average, smaller, they're also making them more nocturnal. A survey, by scientists at University of California, Berkeley, of when mammals are active these days showed that mammals living near people try to avoid them by coming out when the humans have retreated to their homes and/or beds. This observation can be added to the long-known fact that mammals generally try to avoid people by moving away from them in space.

The authors said humans are best viewed as "superpredators", creatures who are abundant and terrifying to many species. Our presence causes wild mammals to always be on guard, constantly pumped full off stress hormones, spending a huge amount of their time in a state of hypervigilance – even when local humans are not bearing arms. Living this way means mammals neglect to undertake basic survival behaviours such as eating and breeding. The study's authors doubted, too, that moving further into the night was a great fix – naturally diurnal animals, they said, would likely struggle to live amongst the nocturnally adapted animals already active then.

Mammals as small creatures who live by night because the day is occupied by a tyrannical foe… it sounds familiar, doesn't it? We humans, just one species, are having a global impact on our mammalian cousins that's akin to the oppressive effect dinosaurs had.

Only we humans operate much quicker than dinosaurs. Damage that accumulated over the course of 125,000 years may be considered short by geological standards, but the effects of what has happened in the last 40–50 years are more like an asteroid crash. In October 2018, the World Wildlife

Fund released its biannual Living Planet Report, which stated that across all vertebrates the average species was 60 per cent less abundant than it was in 1970. This followed a German report from the 'mammals are not unique' category, published in 2017, showing that in the previous 27 years insect populations had fallen by 75 per cent.

Also, in October 2018, Danish researchers asked how long it might take for the mammal species that humans have killed and are killing to be replaced by re-evolved types if humans were to be no more – that is, how many years might it take for evolution to craft mammals like those that humans have wiped out from the species that currently remain. They concluded that for mammalian diversity to recover from human interference would take at least 3–7 million years.

* * *

I saw the fox cubs again the next time we visited Cristina's mum, but the visit after that – no matter how often I crept to the forest's edge – they were gone.

Two months later, in the summer, heading out for a walk, Cristina's mum said absent-mindedly, as she flicked the light switch, 'Let's turn this off and help save the planet.'

'Save the *planet*?' Bella squawked at me bemused. This remark made absolutely no sense to her six-year-old brain – how could something as big and all-consuming as the entire world need saving? And what did something as trivial as turning off a light have to do with it anyway?

This exchange happened when people, post-David Attenborough's *Blue Planet 2*, were being encouraged to use less plastic, and I'd been dismayed that very lunchtime by how much plastic waste I had generated making just one unremarkable meal.

'Imagine a rabbit, Bella,' I began, 'a rabbit eats the grass and other plants that grow where it lives. And it drinks water from puddles and streams near where it lives too.'

'Yes…' she said, clearly unsure about how this related to the whole planet.

'Well, the rabbit then poos on the ground where it lives. And it wees there too. So, really it takes nothing from the planet. It just borrows water and food, to grow and live. Then, when the rabbit eventually dies, a fox or a crow will eat it, and maggots and other bugs will eat whatever's left. Slowly, even its skeleton will rot away. It lives its life and leaves no trace.'

I was about to compare this with my trash-strewn wake, but her attention was gone. She ran off to inspect a hedgerow, leaving me to quietly contemplate how I had that day generated more non-degradable waste than the entire population of wild rabbits that has ever lived.

It was easy to pause the lecture because it felt rather heavy to load all this onto the shoulders of a six-year-old. What had I been building toward anyway? Something like, 'You see, darling, animals are dying, and the climate is moving toward a state that will be catastrophic for human civilisation as we know it. And your generation's task – actually mine and yours, because we probably don't have that much time – is to Save the Planet by being as close to carbon neutral as possible and to desist from constantly destroying natural habitats. This, sweetheart, comes despite the dominant 21st century narrative of your generation being that you can "have it all!" Let's start by not buying that crappy plastic toy you really want because as soon as you were done with it, it would choke a dolphin.'

No child deserves that. But maybe we need them to internalise some of this because these notions don't seem to run deep enough in us grown-ups.

I am writing this – having read the Mammal Society's survey of UK mammals, the WWF's assessment of the world's vertebrates, the study by the meticulous group of German entomologists who directly catalogued insects for 27 years, and all the rest of it – and at some level, I still think surely, it's not *that* bad?

Projections, percentages, numbers, news from elsewhere – none of it is fully penetrative. There remains a disjoint between the data that describes the natural world and my gut feeling about life's resilience – *But those fox cubs*, I think. And what about the scores of seals we went to see this summer? The green from my window right now; the rabbits, squirrels, magpies, ducks and geese in the park across the road and, hey, I've seen rats there too, and, one time, a deer...

Where *Blue Planet* succeeded was in showing its viewers the turtle trapped in a plastic bag and the dead albatross killed by swallowing a plastic toothpick. We hold those bags in our hands and grasp how they could suffocate or strangle an animal. Maybe this is what every documentary should do. Report on how orangutans are smart, then show the footage of a lone male in a flattened forest trying to push a bulldozer away with his bare hands.

People who spend their lives observing wild animals complain that the mass media is uninterested in reporting on a calamitous 80 per cent decrease in population size. They say that the media only come calling when a new extinction is declared – such as the Chinese river dolphin – so that outlet after outlet runs stories asking '*What have we done?*' only when it is too late.

Biologists and conservationists describe our current time as earth's sixth mass extinction. But when you think of mass extinctions – like the one that wiped out the dinosaurs 66 million years ago or the Great Dying that occurred 252 million years ago – you picture a world engulfed in poisonous gases or rumbling with spewing volcanoes or smashed by an asteroid; the planet is littered with corpses, the air thick with the stench of death. It looks nothing like this. There is a failure of comprehension. What does a mass extinction look like? Why, it can look exactly like this: like a feast, a time of absurd plenitude.

One day someone, or something, will write the history of this extinction, just like we try to fathom what happened in the previous five. They will have to reckon with the behaviour of a single unprecedented species, the smartest that's ever been, whose smarts have seen it spread like an algal bloom. What they will say about us is uncertain – will we keep growing until we've smothered and poisoned all about us, like a toxic algal bloom, or will we find ways of living less destructively?

I often think of an online comment below a newspaper report about the global biomass study and humans' having ablated 83 per cent of wild mammals' mass. The comment said, 'Survival of the fittest. Humans rule and others die out because we adapt better. […] Stop the moaning!'

And I think too about the time this autumn when a friend was on a panel that discussed the effects discarded plastic is having in the oceans. Before the panel, I remarked that surely she didn't have to worry too much about there being a significant pro-plastic presence. But two days later, my friend told me the audience was 50:50 on the idea of trying to control plastic waste. 'I get REALLY pissed off when people tell me what to do!', one audience member had barked into the microphone.

And me? Well, I've cut down on plastic, compost whatever I can, recycle more diligently and look at my bin's contents more regretfully. I cry when I watch an orangutan wrestle a bulldozer. I try to eat less meat, but this burger surely won't hurt too much. I fly less than I used to, but I'm looking forward to returning to New York in 2019, it's just one flight, it's just little me, you gotta live, etc., etc. … I could do so much more.

These conflicts and moral struggles and crises of conscience will be lost to future historians. Anyone watching from any distance could only conclude that *Homo sapiens* is killing species at a furious rate, knowingly and because it wants to.

* * *

In December, I return to the forest-facing bench with my morning coffee and huddle against a chilly, damp breeze that from time-to-time blows hard. It's weather that says '*Go inside!*'. Greens are rare now, there are mosses, yes, some brambles and occasional fern fronds, but, no, this is a brown and drawn scene; brown and naked. Now, while the two oaks stand stoic, the birch and hornbeam bend and wave with the wind; their tiny crowns swaying atop long, spindly, sun-chasing trunks – transplanted to a field, these trees would look obscene.

Everything is suspended – the earth's orbit around the sun is written into the DNA of this place. The wood speaks to how, sometimes, barren times should be simply waited out; endured until favour comes again, *'Sleep, why don't you?',* it asks. The hedgehogs sleep, the dormice too, and the bats are tucked up for the winter. Times will be good again, rest, the bluebells will wake you.

I walk to where the fox cubs had played. The leaf-filled mouth of a den is all there is to recall them by. Nothing else says foxes, they've left no trace. I hope, really hope, I get to see more of them this coming spring.

Time here feels cyclic – the forest's gully will soon be a stream again, pools will re-form in the landscape's depressions, winter will do as winter does, and in January and February, the foxes will, one trusts, find each other, knowing somehow that coupling then will mean their young are born just as the first bluebells bloom… Because that's what life does, isn't it? It comes back, racing to catch the sunshine. Only time isn't cyclic, it moves only forward, nothing is guaranteed.

December 2018, East Sussex

Selected Reading

Introduction

Wilson, D. E., Reeder, D. M. (eds) (2005). *Mammal Species of the World: A Taxonomic and Geographic Reference* (3rd edition). Johns Hopkins University Press.

Chapter 1

Chance, M. R. A. (1996). Reason for externalization of the testis of mammals. *Journal of Zoology*, 239: 691–695.

Kleisner, J., Ivell, R., Flegr, J. (2010). The evolutionary history of testicular externalization and the origin of the scrotum. *Journal of Biosciences*, 35: 27–37.

Lovegrove, B. G. (2014). Cool sperm: why some placental mammals have a scrotum. *Journal of Evolutionary Biology*, 27: 801–814.

Moore, C. R. (1926). The biology of the mammalian testis and scrotum. *Quarterly Review of Biology*, 1: 4–50.

Sharma, V., Lehmann, T., Stuckas, H., Funke, L., Hiller, M. (2018). Loss of RXFP2 and INSL3 genes in Afrotheria shows that testicular descent is the ancestral condition in placental mammals. *PLoS Biology*, 16: e2005293.

Chapter 2

Burrell, H. (1927). *The Platypus: Its Discovery, Zoological Position, Form and Characteristics, Habits, Life History etc.* Angus & Robertson (Sydney).

Darwin, C. R. (1845). *Journal of Researches into the Natural History and Geology of the Countries Visited During the Voyage of HMS 'Beagle' Round the World*. John Murray.

Griffiths, M. (1978). *The Biology of the Monotremes*. Academic Press.

Hall, B. K. (1999). The paradoxical platypus. *Bioscience*, 49: 211–218.

Scheich, H., et al. (1986). Electroreception and electrolocation in platypus. *Nature*, 319: 401–402.

Chapter 3

Harper, P. S. (2008). *A Short History of Medical Genetics*. Oxford University Press.

Josso, N. (2008). Professor Alfred Jost: the builder of modern sex differentiation. *Sexual Development*, 2: 55–63.

Morgan, G. J. (1998). Emile Zuckerkandl, Linus Pauling, and the molecular evolutionary clock, 1959–1965. *Journal of the History of Biology*, 31: 155–178.

Rens, W., et al. (2004). Resolution and evolution of the duck-billed platypus karyotype with an X1Y1X2Y2X3Y3X4Y4X5Y5 male sex chromosome constitution. *Proceedings of the National Academy of Sciences of the USA*, 101: 16257–16261.

Sinclair, A. H., et al. (1990). A gene from the human sex-determining region encodes a protein with homology to a conserved DNA-binding motif. *Nature*, 346: 240–244.

Sutton, E., et al. (2010). Identification of SOX3 as an XX male sex reversal gene in mice and humans. *Journal of Clinical Investigation*, 121: 328–341.

Wallis, M. C., Waters, P. D., Graves, J. A. (2008). Sex determination in mammals – before and after the evolution of SRY. *Cellular and Molecular Life Sciences*, 65: 3182–3195.

Chapter 4

Ah-King, M., Barron, A. B., Herberstein, M. E. (2014). Genital evolution: why are females still understudied? *PLoS Biology*, 12: e1001851.

Laurin, M. (2010). *How Vertebrates Left the Water*. University of California Press.

Pough, F. H., Janis, C. M., Heiser, J. B. (2013). *Vertebrate Life* (9th edition). Pearson.

Sanger, T. L., Gredler, M. L., Cohn, M. J. (2015). Resurrecting embryos of the tuatara, *Sphenodon punctatus*, to resolve vertebrate phallus evolution. *Biology Letters*, 11: 20150694.

Shubin, N. H., Daeschler, E. B., Jenkins, F. A. Jr. (2006). The pectoral fin of *Tiktaalik roseae* and the origin of the tetrapod limb. *Nature*, 440: 764–771.

Tschopp, P., et al. (2014). A relative shift in cloacal location repositions external genitalia in amniote evolution. *Nature*, 516: 391–394.

Wagner, G. P., Lynch, V. J. (2005). Molecular evolution of evolutionary novelties: the vagina and uterus of therian mammals. *Journal of Experimental Zoology. Part B, Molecular and Developmental Evolution*, 304: 580–592.

Chapter 5

Carroll, S. B. (2005). *Endless Forms Most Beautiful: The New Science of Evo Devo and the Making of the Animal Kingdom.* Weidenfeld.

Hartman, C. G. (1920). Studies in the development of the opossum *Didelphys virginiana* L. V. The phenomena of parturition. *Anatomical Record*, 19: 251–261.

Nowak, R. M. (2005). *Walker's Marsupials of the World.* Johns Hopkins University Press.

Tyndale-Biscoe, H., Renfree, M. (1987). *Reproductive Physiology of Marsupials.* Cambridge University Press.

Wagner, G. P. (2014). *Homology, Genes, and Evolutionary Innovation.* Princeton University Press.

Weismann, A. (1881). *The Duration of Life.*

Chapter 6

Burton, G. J., Fowden, A. L. (2015). The placenta: a multifaceted, transient organ. *Philosophical Transactions of the Royal Society B, Biological Sciences*, 370: 20140066.

Furness, A. I., et al. (2015). Reproductive mode and the shifting arenas of evolutionary conflict. *Annals of the New York Academy of Sciences*, 1360: 75–100.

Haig, D. (1993). Genetic conflicts in human pregnancy. *Quarterly Review of Biology*, 68: 495–532.

Haig, D. (2015). Q & A. *Current Biology*, 25: R700–702.

Janzen, F. J., Warner, D. A. (2009). Parent–offspring conflict and selection on egg size in turtles. *Journal of Evolutionary Biology*, 22: 2222–2230.

Moore, W. (2005). *The Knife Man: Blood, Body-snatching and the Birth of Modern Surgery.* Bantam.

Pijnenborg, R., Vercruysse, L. (2004). Thomas Huxley and the rat placenta in the early debates on evolution. *Placenta*, 25: 233–237.

Pijnenborg, R., Vercruysse, L. (2013). A. A. W. Hubrecht and the naming of the trophoblast. *Placenta*, 34: 314–319.

Trivers, R. (1974). Parent–offspring conflict. *American Zoologist*, 14: 249–264.

Chapter 7

Blackburn, D. G., Hayssen, V., Murphy, C. J. (1989). The origins of lactation and the evolution of milk: a review with new hypotheses. *Mammal Review*, 19: 1–26.

Daly, M. (1979). Why don't male mammals lactate? *Journal of Theoretical Biology*, 78: 325–345.

Francis, C. M., et al. (1994). Lactation in male fruit bats. *Nature*, 367 691–692.

Lefèvre, C. M., Sharp, J. A., Nicholas, K. R. (2010). Evolution of lactation: ancient origin and extreme adaptations of the lactation system. *Annual Review of Genomics and Human Genetics*, 11: 219–238.

Oftedal, O. T. (2012). The evolution of milk secretion and its ancient origins. *Animal*, 6: 355–368.

Pond, C. M. (1977). The significance of lactation in the evolution of mammals. *Evolution*, 31: 177–199.

Schiebinger, L. (1993). Why mammals are called mammals: gender politics in eighteenth-century natural history. *American Historical Review*, 98: 382–411.

Vorbach, C., Capecchi, M. R., Penninger, J. M. (2006). Evolution of the mammary gland from the innate immune system? *Bioessays*, 28: 606–616.

Chapter 8

Broad, K. D., Curley, J. P., Keverne, E. B. (2006). Mother–infant bonding and the evolution of mammalian social relationships. *Philosophical Transactions of the Royal Society B, Biological Sciences*, 361: 2199–2214.

Clutton-Brock, T. H. (1991). *The Evolution of Parental Care.* Princeton University Press.

Graham, K. L., Burghardt, G. M. (2010). Current perspectives on the biological study of play: signs of progress. *Quarterly Review of Biology*, 85: 393–418.

Lukas, D., Clutton-Brock, T. H. (2013). The evolution of social monogamy in mammals. *Science*, 341: 526–530.

Numan, M. (2007). Motivational systems and the neural circuitry of maternal behavior in the rat. *Developmental Psychobiology*, 49: 12–21.

Pedersen, C. A., Prange, A. J. Jr. (1979). Induction of maternal behavior in virgin rats after intracerebroventricular administration of oxytocin. *Proceedings of the National Academy of Sciences of the USA*, 76: 6661–6665.

Rilling, J. K., Young, L. J. (2014). The biology of mammalian parenting and its effect on offspring social development. *Science*, 345: 771–776.

Spinka, M., Newberry, R. C., Bekoff, M. (2001). Mammalian play: training for the unexpected. *Quarterly Review of Biology*, 76: 141–168.

Zohar, O., Terkel, J. (1991). Acquistion of pine cone stripping behaviour in black rats (*Rattus rattus*). *International Journal of Comparative Psychology*, 5(1): 1–6.

Chapter 9

Archibald, J. D. (2012). Darwin's two competing phylogenetic trees: marsupials as ancestors or sister taxa? *Archives of Natural History*, 39: 217–233.

Close, R. A., et al. (2015). Evidence for a mid-Jurassic adaptive radiation in mammals. *Current Biology*, 25: 2137–2142.

Foley, N. M., Springer, M. S., Teeling, E. C. (2016). Mammal madness: is the mammal tree of life not yet resolved? *Philosophical Transactions of the Royal Society B, Biological Sciences*, 371: 20150140.

Goswami, A. (2012). A dating success story: genomes and fossils converge on placental mammal origins. *EvoDevo*, 3: 18.

Hillenius, W. J. (1992). The evolution of nasal turbinates and mammalian endothermy. *Paleobiology*, 18: 17–29.

Kemp, T. S. (2005). *The Origin and Evolution of Mammals*. Oxford University Press.

Luo, Z-X. (2007). Transformation and diversification in early mammal evolution. *Nature*, 450: 1011–1019.

Madsen, O., et al. (2001). Parallel adaptive radiations in two major clades of placental mammals. *Nature*, 409: 610–614.

Murphy, W. J., et al. (2001). Molecular phylogenetics and the origins of placental mammals. *Nature*, 409: 614–618.

Novacek, M. J. (1992). Mammalian phylogeny: shaking the tree. *Nature*, 356: 121–125.

Simpson, G. G. (1945). The principles of classification and a classification of mammals. *Bulletin of the American Museum of Natural History*, 85: 1–350.

Springer, M. S., et al. (1997). Endemic African mammals shake the phylogenetic tree. *Nature*, 388: 61–64.

Ungar, P. S. (2014). *Teeth: A Very Short Introduction*. Oxford University Press.

Chapter 10

Bennett, A. F. (1991). The evolution of activity capacity. *Journal of Experimental Biology*, 160: 1–23.

Bennett, A. F, Ruben, J. A. (1979). Endothermy and activity in vertebrates. *Science*, 206: 649–654.

Dhouailly, D. (2009). A new scenario for the evolutionary origin of hair, feather, and avian scales. *Journal of Anatomy*, 214: 587–606.

Farmer, C. G. (2000). Parental care: the key to understanding endothermy and other convergent features in birds and mammals. *American Naturalist*, 155: 326–334.

Hayes, J. P., Garland, T. Jr. (1995). The evolution of endothermy: testing the aerobic capacity model. *Evolution*, 49: 836–847.

Huttenlocker, A., Farmer C. G. (2017). Bone microvasculature tracks red blood cell size diminution in Triassic mammal and dinosaur forerunners. *Current Biology*, 27: 48–54.

Kemp, T. S. (2006). The origin of mammalian endothermy: a paradigm for the evolution of complex biological structure. *Zoological Journal of the Linnean Society*, 147: 473–488.

Koteja, P. (2000). Energy assimilation, parental care and the evolution of endothermy. *Proceedings of the Royal Society B, Biological Sciences*, 267: 479–484.

Koteja, P. (2004). The evolution of concepts on the evolution of endothermy in birds and mammals. *Physiological and Biochemical Zoology*, 77: 1043–1050.

Lovegrove, B. G. (2016). A phenology of the evolution of endothermy in birds and mammals. *Biological Reviews*, 92: 1213–1240.

Maderson, P. F. A. (1972). When? Why? And how? Some speculations on the evolution of the vertebrate integument. *American Zoologist*, 12: 159–171.

McNab, B. K. (1978). The evolution of homeothermy in the phylogeny of mammals. *American Naturalist*, 112: 1–21.

Stenn, K. S., Zheng, Y., Parimoo, S. (2008). Phylogeny of the hair follicle: the sebogenic hypothesis. *Journal of Investigative Dermatology*, 128: 1576–1578.

Chapter 11

Allin, E. F. (1975). Evolution of the mammalian middle ear. *Journal of Morphology*, 147: 403–437.

Benni, J. J., et al. (2014). Biogeography of time partitioning in mammals. *Proceedings of the National Academy of Sciences of the USA*, 111: 13727–13732.

Buck, L., Axel, R. (1991). A novel multigene family may encode odorant receptors: a molecular basis for odor recognition. *Cell*, 65: 175–187.

Gerkema, M. P., et al. (2013). The nocturnal bottleneck and the evolution of activity patterns in mammals. *Proceedings of the Royal Society B, Biological Sciences*, 280: 20130508.

Heesy, C. P., Hall, M. I. (2010). The nocturnal bottleneck and the evolution of mammalian vision. *Brain, Behavior and Evolution*, 75: 195–203.

Niimura, Y., Nei, M. (2007). Extensive gains and losses of olfactory receptor genes in mammalian evolution. *PLoS ONE*, 2: e708.

Niimura, Y., Matsui, A., Touhara, K. (2014). Extreme expansion of the olfactory receptor gene repertoire in African elephants and evolutionary dynamics of orthologous gene groups in 13 placental mammals. *Genome Research*, 24: 1485–1496.

Svoboda, K., Sofroniew, N. J. (2015). Whisking. *Current Biology*, 25: R137–140.

Takechi, M., Kuratani, S. (2010). History of studies on mammalian middle ear evolution: a comparative morphological and developmental biology perspective. *Journal of Experimental Zoology. Part B, Molecular and Developmental Evolution*, 314: 417–433.

Chapter 12

Briscoe, S.D., Ragsdale, C.W. (2018). Homology, neocortex, and the evolution of developmental mechanisms. *Science*, 362: 190–193.

Calabrese, A., Woolley, S. M. (2015). Coding principles of the canonical cortical microcircuit in the avian brain. *Proceedings of the National Academy of Sciences of the USA*, 112: 3517–3522.

Dugas-Ford, J., Rowell, J. J., Ragsdale, C. W. (2012). Cell-type homologies and the origins of the neocortex. *Proceedings of the National Academy of Sciences of the USA*, 109: 16974–16979.

Harris, K. D. (2015). Cortical computation in mammals and birds. *Proceedings of the National Academy of Sciences of the USA*, 112: 3184–3185.

Harris, K. D., Shepherd, G. M. (2015). The neocortical circuit: themes and variations. *Nature Neuroscience*, 18: 170–181.

Kaas, J. H. (2011). Neocortex in early mammals and its subsequent variations. *Annals of the New York Academy of Sciences*, 1225: 28–36.

Karten, H. J. (1969). The organization of the avian telencephalon and some speculations on the phylogeny of the amniote telencephalon. *Annals of the New York Academy of Sciences*, 167: 164–179.

Karten, H. J. (2015). Vertebrate brains and evolutionary connectomics: on the origins of the mammalian 'neocortex'. *Philosophical Transactions of the Royal Society B, Biological Sciences*, 370: 20150060.

Northcutt, R. G. (2002). Understanding vertebrate brain evolution. *Integrative and Comparative Biology*, 42: 743–756.

Romer, A. S. (1933). *Man and the Vertebrates*. University of Chicago Press.

Rowe, T. B., Macrini, T. E., Luo, Z. X. (2011). Fossil evidence on origin of the mammalian brain. *Science*, 332: 955–957.

Striedter, G. F. (2004). *Brain Evolution*. Sinauer.

Chapter 13

Darwin, C., ed. Darwin, F. (1958). *Selected Letters on Evolution and Origin of Species (With an Autobiographical Chapter)*. Dover Publications.

Estrada, A., et al. (2017). Impending extinction crisis of the world's primates: why primates matter. *Science Advances*, 3: e1600946.

Kaas, J. H. (2013). The evolution of brains from early mammals to humans. *Wiley Interdisciplinary Reviews: Cognitive Sciences*, 4: 33–45.

Kemp, T. S. (2016). *The Origin of Higher Taxa: Palaeobiological, Developmental and Ecological Perspectives*. Oxford University Press and University of Chicago Press.

Martin, R. D. (2012). Primates. *Current Biology*, 22: R785–790.

Young, J. Z. (1950). *The Life of Vertebrates*. Oxford University Press.

Van Wyhe, J., Pallen, M. J. (2012) The 'Annie hypothesis': did the death of his daughter cause Darwin to 'give up Christianity'? *Centaurus* 54; 105–123.

Acknowledgements

Thank you to the unknown footballer who, in 2011, failed to score past me. I hated you for half an hour, but I'm glad this book exists. Thank you to Laura Helmuth, then at *Slate*, for publishing my article on the natural history of externalised testicles. And a huge thank you to Julie Bailey for reading that article, inviting me to write a book and believing that I *could*. Also at Bloomsbury, Anna MacDiarmid has been brilliant to work with; I thank her for all her hard work on making this a reality. It's been great to be a part of Jim Martin's community of Sigma authors and to work with him. My gratitude goes to my editor Catherine Best, and to Julie again, who each made this a better book than the manuscript I submitted.

In proposing this project, I wildly underestimated the magnitude of the task I was setting myself, but I have very fortunately and gratefully received the support of many people. Early on, I remarked to a friend, 'I hope I haven't bitten off more than I can chew.' The response sent me into a spiral of doubt: 'Do only mammals chew?' Little did I know.

I thank the curators and trustees of the American Natural History Museum in New York, the Natural History Museum in London, the Oxford University Museum of Natural History, the British Museum and Down House, for creating such inspiring places to visit.

I thank Tom Kemp, Jenny Graves, Roger Close, Thom Sanger, Patrick Tschopp, Ana Calabrese, Jon Kaas, Caroline Pond and Harvey Karten for discussing their work with me, and, in some instances, for providing feedback on drafts of the manuscript. So many scientists have done so much remarkable work on the topics covered in this book (and I have merely skimmed the surface), my humblest gratitude extends to every one of them. I accept full responsibility for any and all errors that remain in this work.

I'm especially grateful to Helen Scales, whose calm voice frequently guided me. And to Derek, who helped me through the toughest segment – this is the note I struck.

Numerous people read and commented on chapters, and I thank Helen, Bonnie Walker, Emma Bryce, Debra O'Sullivan, Katie Greenwood-Skinner, Eleanor Gould, Emma Steven, Curtis Asante, Jaime McCutcheon, Damian Pattinson and other members of NeuWrite London for doing so.

I know this is no Oscars speech, but I would like to thank my parents – I had no idea this story would be *so* much about bringing up offspring: it has only deepened my gratitude to you. Thanks to my bro too, just because. My sincerest thanks go to my ma-in-law, Susan Castillo Street: she has been unflagging in her championing of this project, the most diligent of readers and simply an integral part of its completion. I now know why she has such a loyal and appreciative cadre of former students.

Cliff Hopkinson taught me what it means to try to write. The term mentor is used far too liberally these days. I reserve it strictly for someone who, with kindness, wisdom and substantial personal investment, helps a novice better him- or herself; someone whose effect is profound. I shouldn't get too serious, we did laugh and drink the whole way through, but, Cliff, you were that mentor. Thank you.

To Isabella and Mariana, you made me better. You daily make me happy. Watching you grow, helping where I can, humbles me. May you both soar. And to Cristina: first, there's the book-specific thank you for tolerating my chaos and mess. Then, there's the larger one; the one for being you, for making us and for loving me. Thank you. I hope that what the three of you mean to me came through in the foregoing; from now on, I'll stick to more conventional expressions of love.

Index